Rush Shippen Huidekoper

Age of the Domestic Animals

Rush Shippen Huidekoper

Age of the Domestic Animals

ISBN/EAN: 9783337401818

Printed in Europe, USA, Canada, Australia, Japan

Cover: Foto ©berggeist007 / pixelio.de

More available books at **www.hansebooks.com**

AGE OF THE DOMESTIC ANIMALS

BEING A COMPLETE TREATISE ON THE

DENTITION OF THE HORSE, OX, SHEEP, HOG, AND DOG,

AND ON THE VARIOUS OTHER MEANS OF DETERMINING THE AGE OF THESE ANIMALS.

BY

RUSH SHIPPEN HUIDEKOPER, M.D.,

VETERINARIAN (ALFORT, FRANCE); PROFESSOR OF SANITARY MEDICINE AND VETERINARY JURISPRUDENCE, AMERICAN
VETERINARY COLLEGE, NEW YORK; LIEUTENANT-COLONEL AND SURGEON-IN-CHIEF NATIONAL GUARD
OF PENNSYLVANIA; FELLOW OF THE COLLEGE OF PHYSICIANS, PHILADELPHIA;
HONORARY FELLOW OF THE ROYAL COLLEGE OF VETERINARY SURGEONS,
LONDON; LATE DEAN OF THE VETERINARY DEPARTMENT,
UNIVERSITY OF PENNSYLVANIA, ETC.

ILLUSTRATED WITH 200 ENGRAVINGS.

PHILADELPHIA AND LONDON:
F. A. DAVIS, PUBLISHER,
1891.

Philadelphia, Pa., U. S. A.
The Medical Bulletin Printing House,
1231 Filbert Street.

TO THE MEMORY OF

ARMAND GOUBAUX,

LATE DIRECTOR OF THE VETERINARY
SCHOOL AT ALFORT,

AND TO

GUSTAVE BARRIER,

PROFESSOR OF ANATOMY AND EXTERIOR AT THE VETERINARY
SCHOOL AT ALFORT,

THIS WORK IS RESPECTFULLY DEDICATED

IN RECOGNITION OF THEIR SCIENTIFIC TEACHINGS, FROM WHICH
I HAVE LARGELY PROFITED AND DRAWN, AND IN SINCERE
APPRECIATION OF THE MANY COURTESIES I
RECEIVED FROM THEM WHEN A
STUDENT AT ALFORT.

THE AUTHOR.

PREFACE.

As an owner of horses and dogs, and later as a student, practitioner, and teacher of veterinary medicine, the author has constantly regretted the poverty of practical books on the domestic animals, and has felt the want of accurate text-books on many of the ordinary subjects concerning them. Upon the age of the domestic animals there exist in English but incomplete and scattered writings; the best of them by writers of the first half of the century, which have only been supplemented in later years by short compends, or the publication of charts of variable merit, with but a small amount of descriptive matter.

The author has attempted to prepare such a book as he feels would have been of interest and service to himself in his association with animals as a layman, and would have aided his studies and appreciation of the anatomy of the teeth, dentition, and the means of determining the age. He hopes, also, that this work will furnish, to students and veterinarians, knowledge which will aid in surgical operations on the mouth.

In French, German, and Italian the literature upon this subject is much more complete, and the articles in these languages are much better illustrated; so that the author has drawn freely in quotations and translations from the list of writers found in the appended Bibliography, and has used many illustrations where he has not believed that new drawings would be an improvement upon the older ones. He wishes to recognize especially his indebtedness for many paragraphs which have been freely translated, and for a majority of the illustrations on the horse's age, from Goubaux and Barrier; for illustrations on the dog, from Girard; to text and the use of illustrations on the hog, from Dr. Olof Schwartzkopff; and the use of illustrations on the human mouth, from the White Dental Manufacturing Company, of Philadelphia.

RUSH S. HUIDEKOPER.

BROADWAY, COR. FIFTY-THIRD STREET,
     NEW YORK, October, 1891.

TABLE OF CONTENTS.

	PAGE
PREFACE, .	v
INTRODUCTION, ...	1
Definition of age; periods of age.	
THE TEETH,	7
DENTITION IN THE HORSE.	9
Incisors: Temporary, permanent, structure, direction. Molars: Temporary, permanent, development and structure.	
ERUPTION OF THE TEETH. .	43
The incisors; tusks; molars.	
DURATION OF THE LIFE OF THE HORSE, ...	46
Principles of examination for determining the age of the horse; characters furnished by the teeth.	
FIRST PERIOD.	53
Eruption of the incisors of first dentition; at birth, one week, one month, three months, four months, five months, six months, ten months.	
SECOND PERIOD.	64
Leveling, progressive use, and falling out of the incisors of the first dentition; one year, sixteen months, twenty months, two years.	
THIRD PERIOD, ...	74
Eruption of the permanent or adult teeth: two and a half years, rising three years, three years, three years off, rising four years, four years, four years off, rising five years, five years, five years off.	
FOURTH PERIOD.	92
Leveling of the permanent incisors; at six years, at seven years, at eight years	
FIFTH PERIOD.	100
Wearing away of the crowns: at nine years, at ten years, at eleven years, at twelve years, at thirteen years, at fourteen years, at fifteen years, at sixteen years, at seventeen years, after nineteen.	

(vii)

PAGE

IRREGULARITIES OF THE DENTAL SYSTEM, ... 122

Irregularities in number—augmentation; diminution; irregularities of form; irregularities by uniting of two incisors; irregularities in form of the dental cup—fissure; irregularities in depth of the dental cups; excess of hardness of the teeth; excess of length of the dental cup; fault of length or size of one jaw; excess of width of upper incisive arch; irregularities by excess or fault of use; excess of length of jaws; excess of length of incisors; excessive length of single teeth; deficiency of length of incisors; deficiency of length of crown of molars; excess of length of crown of molars; accidental irregularities; irregularities from cribbing; artificial irregularities; removal of temporary incisors; bishoping; dressed mouths.

AGE OF THE ASS, MULE, AND HINNY, ... 157

AGE OF THE OX, ... 158

Dentition; incisors, molars; table of eruption. Determination of age by the teeth; eruption of temporary teeth; wearing of the temporary teeth; eruption of permanent teeth; leveling of permanent teeth; wearing away of the crowns. Determination of age by the horns.

AGE OF THE SHEEP AND GOAT, ... 175

Dentition; incisors, molars; eruption of temporary teeth; alteration during the first year; eruption of permanent teeth. Determination of age by the horns.

AGE OF THE HOG, ... 184

Dentition; incisors, tusks, molars. Determination of age by the teeth; eruption of temporary teeth; eruption or permanent teeth; table of dentition.

AGE OF THE DOG, ... 194

Dentition; incisors, tusks, molars. Determination of age by the teeth; eruption of temporary teeth; eruption of permanent teeth. Determination of age from other signs.

AGE OF MAN, ... 203

Dentition, temporary, permanent; determination of age by the teeth.

BIBLIOGRAPHY, 211

INDEX, ... 213

Age of the Domestic Animals.

INTRODUCTION.

AGE is defined by Webster as "the whole duration of a being," or "that part of the duration of a being which is between its beginning and any given time." It is not to be confounded with the term as applied to the various *periods* or epochs of an animal's life, as the *juvenile*, *adult*, and *senile*, which correspond to organic metamorphosis of distinct character and to a marked change in the degree of functional activity. The age of the domesticated animals is a matter of great importance in agricultural commerce, as, in the limited period during which each of them is individually useful, a comparatively short time diminishes greatly the extent of usefulness to which each can be put, and consequently lowers its value as an investment. The means which enable us to judge the age of an animal are based upon certain anatomical and physiological changes which occur in the course of the development of the newborn to the adult animal, and in the deterioration of it from its period of perfection to the decrepitude of its last years. These changes are constant, and, while the animal retains its general form and the tissues are chemically identical, there is a constant renewal of the molecules, which is rapid early in life and slower later. Age can be divided into three periods: First, *juvenile, or the period of growth*, which extends from the birth of the animal to its full development, during which it is gaining in size, in strength, and in intelligence, and is constantly increasing in its qualities and value. Second, *adult, or the stationary period*; in this the animal is at its best; it has every organ complete in size and perfect in function, working in perfect harmony with the rest of the body; its intellect has the most complete control over its perfectly organized muscles, digestive tract, and other

1

(1)

apparatuses, and it is capable of its best and greatest work; in this stage there is an equilibrium established between the waste and repair, and the activity of composition is equal to that of decomposition. Third, *senility, or old age*, the period of deterioration; during this period the animal may have the strength of previous years, it may be capable of as great efforts, and from experience may even display greater intelligence; for specific purposes the animal may at this time be capable of rendering even greater service than it could have given when younger, but its efforts now demand a waste of tissues no longer capable of rapid repair, and continual drains on its organs alter them in size and depreciate them in function, and the whole animal economy undergoes a steady deterioration in usefulness and value.

These periods may be called that of *growth*, the *stationary period*, and that of *decline*.

The veterinarian is the expert referee in all questions of age, and any addition to our knowledge in regard to it needs no apology.

If animals lived a perfectly equable life, the changes which take place in their form and character would be regular and easy to appreciate; but, from overwork, from the alterations produced by excessive food, want of food, or irritant substances; from the effects of variable surroundings, at times trying on the constitution, at other times rendering the external tissues delicate and sensitive to exposure, the wear and tear on them is irregular. Excessive work in one animal and want of food in another open easy avenues for the ravages of time, which would not show on one which was better nourished and better cared for. The well-bred colt training over fast miles from a yearling, and the great draught-horse hauling a load of several tons in our large cities, demand excess of food for their herculean tasks. The greater quantity of food overtaxes the organs in proportion, and, in the same number of years, one animal may live double that of another in usefulness and functional activity.

We judge the age of an animal from its general aspect, from the various changes in the conformation of the body, both external and internal, and from the functional activity of its various organs. All of these are to be examined in detail, and the synthesis of the result usually gives us an accurate indication of its age. The most important details in judging of the age, especially in the horse, are found in the teeth. We find the wear and tear of these organs directly in relation to the amount of aliment which they have had to handle, and the amount of aliment directly in relation to the number of years which the animal has lived; taking always into consideration that the energetic draught-horse and the rapid race-horse, which are obliged to do excessive work, will have their organs more rapidly modified by the excessive nature of the work which they are obliged to perform.

In the ox and sheep, again, the teeth are of great value in judging of age, but the epidermic products in the shape of the horns are also important factors after they have become adult. In the dog, the teeth are valuable in indicating the age of young animals, but in older ones we find that the variable mode of life to which these animals are subject so modify these organs that they are of little value, and in them we must look to alteration of the epidermic appendages of the skin, and to conditions in their general aspect, for our best guide.

In very young animals there is a decided disproportion between the length of the legs and the size of the belly and the rest of the body, which varies with the species, the habits, and aptitudes of the animal. In the herbivora, where the young follow their mothers to pasture and suck the udders standing, the legs are long and capable of holding the animal immediately after birth; the young see well and have the intelligence to avoid ordinary dangers from the first day. In the carnivora, on the contrary, whose search for food is often at a great distance, and so fatiguing that the young could not follow the mother, the young are born with legs too short for use and with closed eyes,

4

which force them to remain quietly in their nests while their mothers are away, and until they have acquired considerable force and strength. In all species the young are born with a short down over the body, which serves to protect them from cold and wet; if the animals are pigmented, this is lighter in color than the coat will be later. The brain and head are always large, and overhang the little faces. The digestive tract is well developed, and frequently gives the belly a swelled appearance. The tissues are soft; the bones, scarcely formed, are flexible; and while, on the one hand, they may favor deviations and deformity, on the other they allow of considerable displacement in the helpless animal without injury to it; the liver is enormous, from its activity in intra-uterine life; the lymphatic ganglia are large, as are the thymus glands and supra-renal capsules; the ventricular portion of the heart and the arteries exceed the veins in size; the lungs are well developed, the kidneys the same, and the nervous system is relatively excessive in size, while the organs of generation may be almost rudimentary.

The functions of the organs are in co-relation with their development. That of digestion is most prominent, and, indeed, is the first and only one in the very young animal, for which it uses its intelligence to direct voluntarily, searching and crying for food almost as soon as born, to find, at the teats of the mother, the nutrition which was interrupted at the rupture of the umbilical cord. Digestion and assimilation of food take place rapidly; the circulation of the blood is frequent; the secretion of saliva, pancreatic fluid, and bile is active, while the excretion of waste matter is small. The rapid chemical changes in the tissues of the very young is productive of a high body temperature.

The function of relation is somewhat limited in the young herbivora, which, however, see, can guide themselves, and move considerable distances; the sense of hearing and taste is developed, while that of smell is almost absent. In the young carnivora, on the contrary, the function of relation is limited at first to the search for and sucking of the mother's milk.

As the animal gets older its parts become more harmonious. In the herbivora the body grows in all directions, the legs become larger without lengthening proportionately, the head elongates, the chest develops, and the belly recedes, until the young has attained the form of the adult animal. This development can be guided, hastened, and aided very materially by judicious feeding and care on the part of man. In the young carnivora the legs elongate after the other changes in development have commenced.

As the animal grows the large intestine increases relatively in size, the liver and kidneys diminish, the venous system increases, the nervous system develops relatively much less, while the generative organs and the special senses become perfect with complete development of all the organs, and the establishment of their perfect functional activity. The points of conformation and character which distinguish the male from the female become marked, while in the very young they were scarcely appreciable.

In the *adult* or *stationary period*, the animals show the distinctive characters of their species, race, family, and individuality. The bodies are harmonious, the bones are solidified, the organs are capable of auxiliary work, and may either be used to produce force or as store-houses of latent force, in the form of flesh, to be consumed and converted into force at a later time. At this period the sexual organs are at their best, and the sexual desire is most active.

In the *senile* or *period of decline*, the rounded lines of the beautiful adult animal disappear, the bony points of the skeleton become prominent, the back drops down, the skin becomes dry and loses its elasticity, the hairs turn white, the head becomes long and pointed, the eyes sink, the muscles atrophy, the tendons contract, and the bones are subject to inflammations. The organs lessen in activity until they are

6 AGE OF THE DOMESTIC ANIMALS.

unable to supply the tissues with nutritive matter equal to that which is destroyed and thrown off, and the animal emaciates. The functions slow; the senses are diminished and finally become extinct.

As the determination of age, in all of the domestic animals, depends primarily on the alterations which take place in their teeth, and, secondarily, upon other signs and changes, the subject will be taken up in regular order: First, the study of the dentition of the animal, the number and structure of its teeth, and the manner of eruption and the alterations which they undergo by use; then the various other organs which undergo characteristic changes with age will be described,—the horns and other epidermic products, the alterations of form and of expression.

The species of animal will be taken up according to its importance and value as an article of commerce.

THE TEETH.

THE teeth are described by Cuivier as "mechanical instruments in the vertebrated animals, at the entrance of the alimentary canal, designed to seize, cut, tear, bruise, and grind nutritive substances, before their transmission to the mouth, pharynx, and the œsophagus."

Professor Owen says: "They present many varieties as to number, size, form, structure, position, and mode of attachment, but are principally adapted for seizing, tearing, dividing, pounding, and grinding the food. In some species they are modified to serve as formidable weapons of offense and defense; in others as aids in locomotion, means of anchorage, instruments for uprooting or cutting down trees, or for transport and working of building materials."

Chauveau says: "Identical in all of our domesticated animals, by their general disposition, their mode of development, and their structure, in their external conformation these organs present notable differences the study of which offers the greatest interest to the naturalist, for it is on the form of its teeth that an animal depends for its alimentation. It is the *régime*, in its turn, which dominates the instincts and commands the diverse modifications in the apparatus of the economy; and there results from this law of harmony so striking a correlation between the arrangement of the teeth and the conformation of the other organs, that an anatomist may truly say: "Give me the tooth of an animal, and I will tell you its habits and its structure."

All of the domestic animals have two sets of teeth: first, those of *first dentition*, which appear at or soon after birth, which are known as *fœtal, temporary, milk, deciduous*, or *caducous* teeth; second, those of the *second dentition*, which are known as *replacing, persistent*, or *permanent* teeth.

(7)

The teeth of both dentitions are placed in each jaw so as to form more or less of an arch, or horseshoe, the convex portion forward and the open end facing toward the posterior part of the mouth and throat. These arches are subdivided into three portions, an anterior, a lateral, and an intermediate, and in each section the teeth have characters which distinguish them from the others, and distinctive forms which indicate the use to which they are adapted.

In the anterior arch the teeth are sharp-edged, and are adapted for cutting or nipping the food; these are called *incisors* (*incidere*, to cut), and are most characteristic in the herbivora.

In the intermediate section the teeth are large, pointed, and sharp, destined to tear the food or to serve as weapons of offense and defense; these are called *tusks, tushes, fangs,* or *canine* teeth; they are most developed in the carnivora, which tear their food.

The lateral arches have broad, strong teeth, with flat surfaces, admirably fashioned for the grinding of the food, from which they take the name of *molars* (*molere*, to grind), or grinders. These are sometimes subdivided into *premolars* and *post-molars*, or molars proper.

As the teeth on the two sides are symmetrical, for convenience of description zoölogists have adopted the following formula to indicate the number of teeth in either dentition or at any given period of an animal's life; it gives only the teeth on one side of the mouth:—

I = incisors, T = tusks. M = molars.

Thus, for the permanent dentition of the horse:—

$$\text{Superior jaw } \atop \text{Inferior jaw }\} \quad I\ {3 \atop 3},\ T\ {1 \atop 1},\ M\ {6 \atop 6},\ = 20; \text{ both sides, 40.}$$

Or, more simply:—

$$\frac{3\ .\ 1\ .\ 6}{3\ .\ 1\ .\ 6} = 40.$$

Formula
$$\left\{\begin{array}{l}\text{Temporary,} \\ \\ \text{Permanent..}\end{array}\right.$$
$$\cdot \;\; \frac{3 \,.\, 0 \,.\, 3}{3 \,.\, 0 \,.\, 3} = 24$$
$$\frac{3 \,.\, 1 \,.\, 6}{3 \,.\, 1 \,.\, 6} = 40$$

THE horse has thirty-six or forty teeth, according to the sex of the animal, which are divided into three groups,—the incisors, the molars, and the tusks. In front are the incisors, to each side the tush or canine teeth, and still farther back, on the sides, the molars. (Fig. 1.) The first are used to grasp and cut the food, the second to tear it, and the third to bruise and grind it up. Following their position in the jaw, the teeth form a parabolic curve, known as the dental arches, of which there are two,—one in the upper jaw and one in the lower jaw. These arches are again subdivided into three portions, an anterior and two lateral. The incisor teeth form the anterior part, in the shape of a half-circle, convex in front. On each side, directly behind the incisors is a space, larger or smaller, according to the sex of the animal, which corresponds to the intermaxillary and maxillary bones, and extends to the lateral part of the dental arch. This is known as the interdental space. It is plain and extensive in mares, because they ordinarily do not have tush teeth; but when they do, as in the male, it is divided into two parts, known as the anterior and posterior. The latter, in the lower jaw, is known as the bar of the jaw. Further behind, to the right and left, forming, as it were, branches or sides to the dental arch, are found the molar teeth.

In the adult animal there are in each jaw six incisors, two tush teeth, and twelve molars, making a total of forty for the horse, and, without the tush teeth, thirty-six for the mare. In the colt there are twelve incisors and twelve molars, the latter divided into rows of three above and three below on each side. In the young animal the tush teeth do not exist. How-

(9)

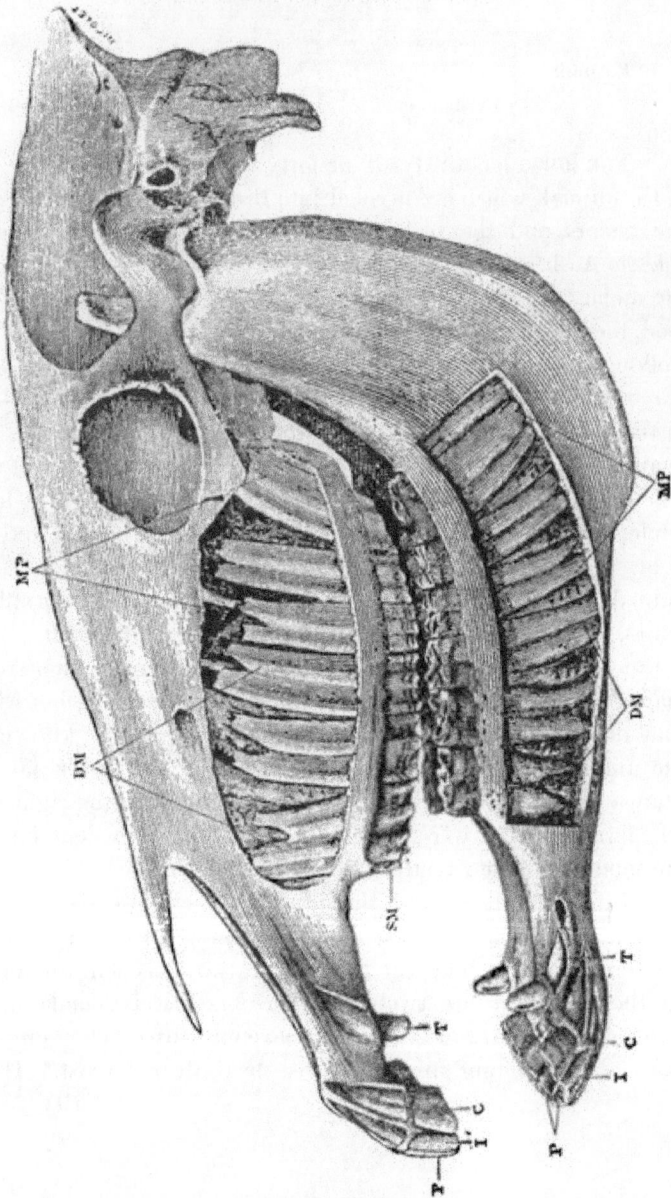

FIG. 1.

Complete dentition of the horse. P, pincher teeth ; I, intermediate teeth ;
C, corner teeth ; T, tush or canine teeth ; S M, supplementary premolars ; D M,
deciduous molars ; M P, permanent molars.

ever, in the place where they will appear later, we sometimes
find small, rudimentary teeth with no defined shape. We again
sometimes find, both in the young and in the adult animal,
a more or less rudimentary premolar tooth, commonly known
as the *wolf tooth*, which raises the number in the adult animal
to forty-four.

INCISORS.

INCISORS OF FIRST DENTITION.

The milk-teeth, known also as deciduous or fœtal teeth, are
twelve in number,—six in each jaw, three on each side. The

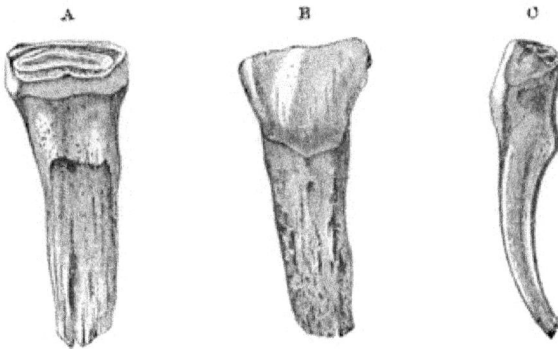

FIG. 2.

A milk-tooth. A, posterior face; B, anterior face; C, profile.

middle ones are known as pincher teeth, the next as the intermediate teeth, and the outside ones as the corner teeth. They
are at first imbedded in the body of the bone, and covered by
the gum; but when they have protruded from the alveolar
cavities they form a half-circle, convex in front. Compared
with the permanent teeth, they are shorter; they have a constriction in the centre which is known as the neck, which
divides them into a free portion or crown, and an imbedded portion or root (Fig. 2). They are dead white, milky, or yellowish
white in color. The anterior face of these teeth is convex in
both directions and roughened by little parallel, longitudinal
ridges and depressions, which, however, become worn off, and

have a smooth and polished surface as the animal becomes older. The posterior face is concave from above to below, and slightly convex from side to side. The internal border of each tooth is thicker than the external. In a virgin tooth (Fig. 3), or one which is not worn by use, the free portion of crown is divided from in front to behind and limited by two borders, one anterior, *a*, and the other posterior, *b*, which are separated by a cavity known as the cup, *c*. The anterior border is the highest and longest; it is shaped convex transversely, and is the first portion of the tooth to come through the gum. The posterior border is

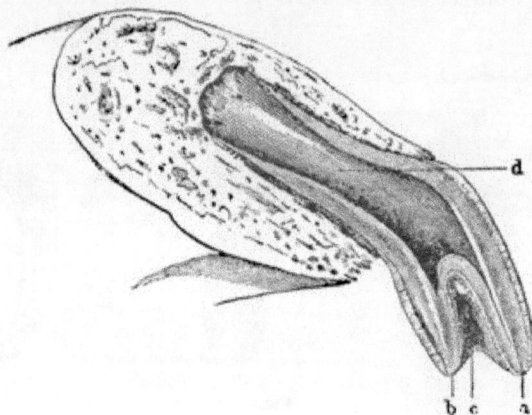

FIG. 3.
Longitudinal section of a virgin milk-tooth in its alveolar cavity.

shorter and appears later, but soon reaches, however, the level of the anterior border from the wearing down of the latter. From the same cause the cavity, or cup, which first existed, gradually disappears. This is known as the leveling of the teeth. The imbedded portion or root also contains a cavity, known as the internal dental cavity, or pulp-cavity (*d*), which protects the papilla or pulp of the tooth; but as the animal becomes older the tooth elongates by the growth of its imbedded portion, and the internal cavity diminishes in calibre and is nearly obliterated by a deposit of bony substance; at the same

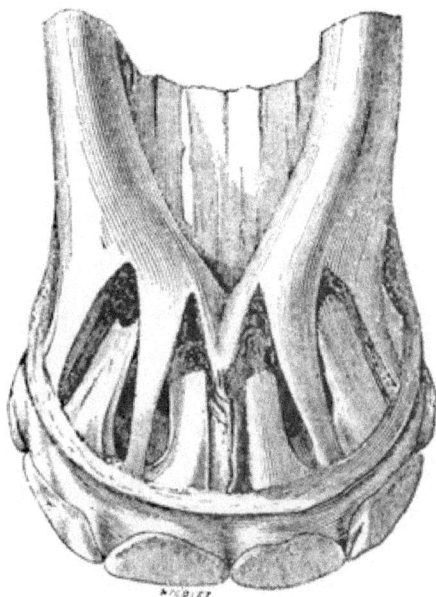

FIG. 4, a.

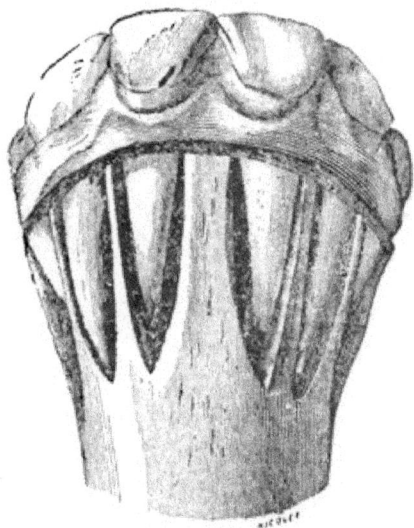

FIG. 4, b.

Temporary incisors, front face. a, upper jaw; b, lower jaw. The roots are uncovered to show their position in the alveolar cavities.

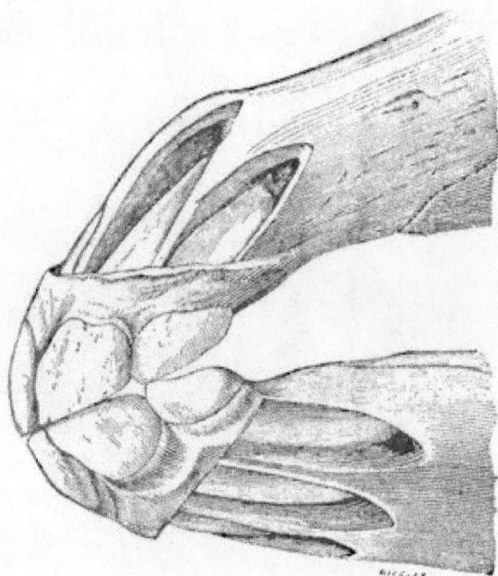

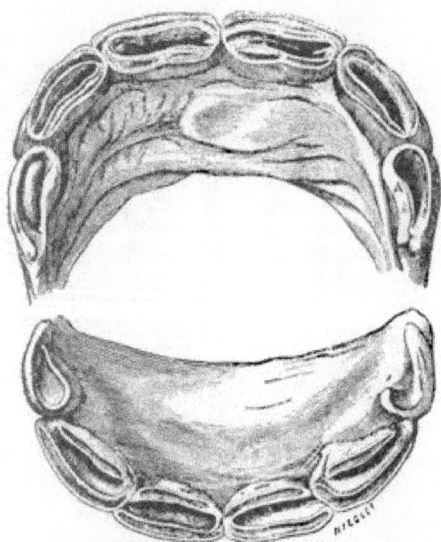

FIG. 5.

Temporary incisors, profile and table. In profile the roots are uncovered to show
their position in their alveolar cavities.

time the bone of the jaw increases in size, and the permanent
incisors form in them.

These develop behind the deciduous teeth, and are at first
separated from them by a bony septum, which, however, usually
becomes thinned, and the permanent tooth pushes the other out.

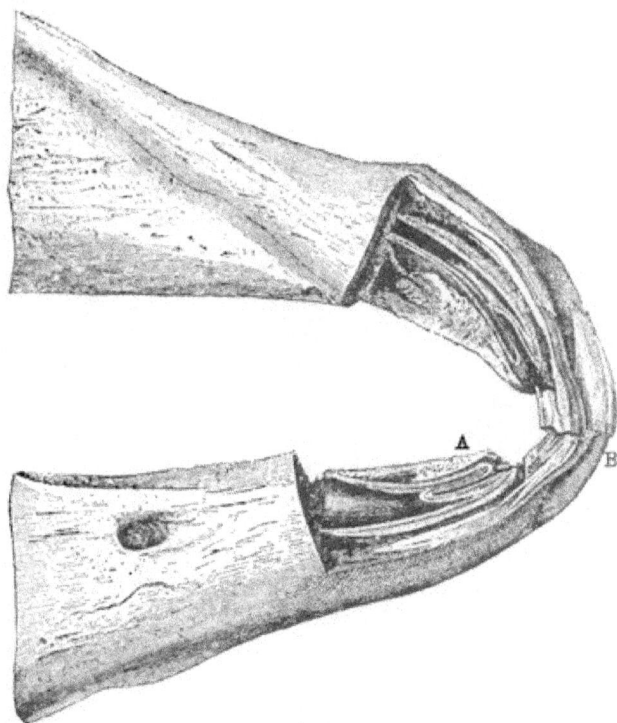

FIG. 6.
A, permanent incisor just before eruption; B, deciduous incisor in place.

but sometimes they come from an independent opening and
the milk-tooth remains in place.

The absorption of the bone, by pressure of the developing
teeth, is frequently attended by constitutional phenomena, loss of
appetite, sluggishness, etc., and I have seen it produce convulsions in a colt two and a half years old.

There is an old superstition that the period of eruption of

the permanent incisors predisposes the animal to the contraction of contagious fevers.

PERMANENT INCISORS.

The permanent incisors are six in number in each jaw, like the deciduous incisors which they replace; they, like the latter, are known as the pincher teeth, intermediate teeth, and corner teeth. They are longer than the milk-teeth, are more of a wedge-shape, have no constriction or neck separating the crown from the root, and are of a bluish-white, instead of the cream color of the others. The permanent incisors (Fig. 7) are shaped somewhat

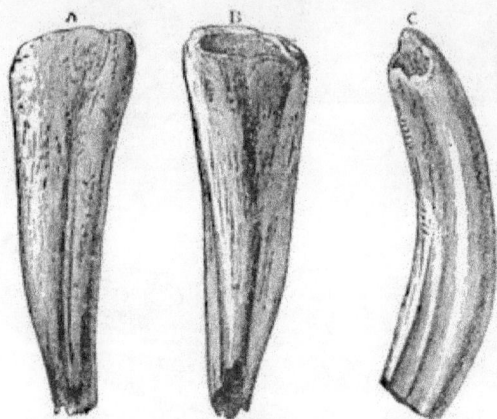

FIG. 7.

A virgin inferior permanent incisor. A, anterior face; B, posterior face; C, profile.

like an irregular cone, of which the base corresponds to the crown or free extremity, and the summit to the root or the imbedded extremity. They are curved on their long axis, with their convex face in front and their concave face behind, and they are also somewhat twisted on their long axis; the pincher teeth are slightly so, the intermediate teeth more so, and the corner teeth sometimes make almost a decided S. The free portion, or crown, is flattened from in front to behind, at the level of the gums the two axes have about the same diameter, and the root or imbedded portion is flattened from side to side.

The anterior face, which is widest above, is flat from side to side and convex from above to below. It frequently is grooved by a little canal, which is most distinct on the free portion of the tooth. The incisors of the upper jaw are more convex than those of the lower jaw, and they frequently have two little longitudinal canals instead of one. The posterior face is convex and rounded from side to side and concave from above to below. The internal border of the tooth is thicker than the external; this is more marked in the imbedded portion than in

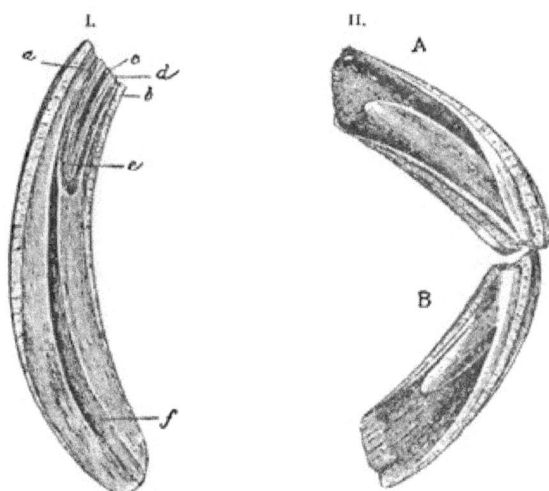

Fig. 8.

I. Longitudinal section of a permanent incisor, leveled. II. Eccentric and longitudinal section of superior and inferior intermediate permanent incisor, to show the differential characters of the dental cup.

the free portion. The free portion, or crown, is flattened from in front to behind, and is hollowed out by a cavity known as the external dental cavity or *cup.* (Fig. 7. *B* and *C.*) In the virgin tooth there is a border in front and a border behind ; the latter is not so high as the first, but in old teeth these become worn off until perfectly level, when the extremity of the tooth takes the name of the *table* of the tooth. (Fig. 8. I, upper extremity.)

The external dental cavity (Fig. 9. *a a a*), surrounded by

2

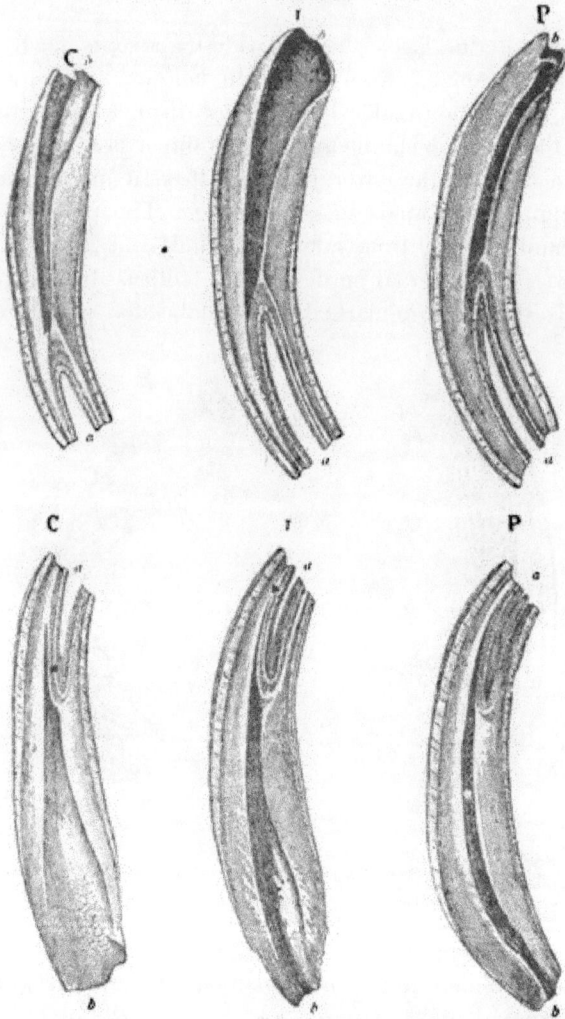

FIG. 9.

Longitudinal and antero-posterior sections of permanent incisors. P, pinchers; I, intermediate; C, corner; a a a, free extremities; b b b, roots.

the cup of enamel, has an irregular, conical form, the base toward the free extremity of the tooth itself and the point imbedded in the tooth toward its root, but inclined slightly toward the posterior border, especially in the inferior tooth.

This is sometimes a hollow cup, but it is frequently filled with cement, making the free extremity, even in the virgin tooth, in such cases perfectly solid. The cups in the inferior teeth are less deep than those in the superior incisors.

In the incisors of the lower jaw the pincher teeth have a cup with an average depth of 16 to 18 mm.; the cup of intermediate teeth is 18 to 20 mm., and that of the corner teeth is 11 mm. to 13 mm. In the upper jaw the cup of the pinchers is 25 mm. to 27 mm. deep; that of the intermediate teeth, 27 mm. to 28 mm. deep; and that of the corner teeth somewhat less, 18 mm. to 20 mm. (Fig. 9, *a a a ;* Fig. 12, *c.*) The cups in the lower teeth incline nearer the posterior border of the tooth than in the upper ones.

In the imbedded portion of the tooth there is a large canal which holds the dental pulp or papilla, which is known as the *internal dental cavity.* (Fig. 9, *b ;* Fig. 12, *P.*) Examine Fig. 9, and we find that this cavity occupies the centre of the tooth, in the root, but toward the crown inclines toward the anterior face and penetrates between the latter and the cup of enamel. As the tooth becomes older this cavity becomes filled with an ivory-like substance, softer and darker colored than the rest of the tooth, and when it appears on the table of the tooth, from the wearing away of the crown, it is known as the *"dental star."*

In order to understand clearly the various forms which the table of the tooth assumes as it wears away, a series of sections are shown in Fig. 10, representing the table as it appears from the use of successive years.

A. The rubbing surface of the dental table is at first flattened from in front to behind; that is to say, its transverse diameter is greater than the antero-posterior. (Sections 1 and 2.)

B. It becomes oval; there is still a disproportion between the extent of its two diameters, but the transverse diameter remains greater than the antero-posterior. (Sections 3, 4, and 5.)

C. It becomes rounded and its two diameters are nearly equal. (Sections 6 and 7.)

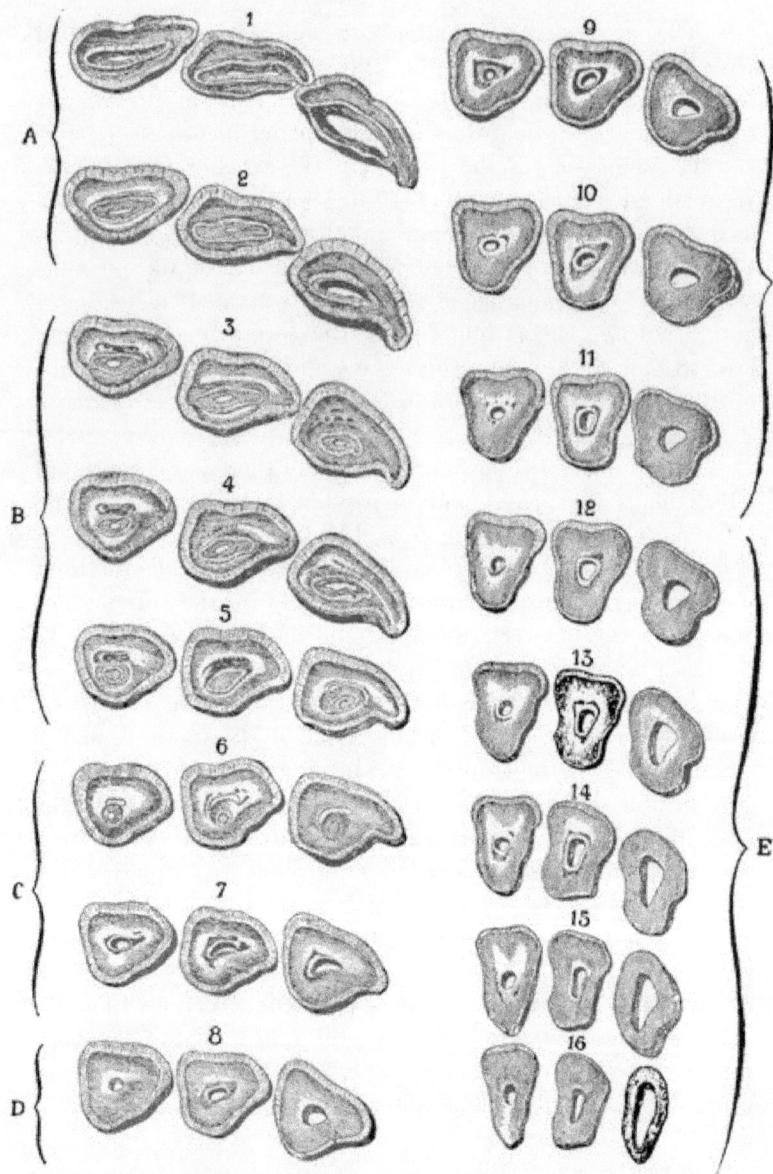

FIG. 10.

Series of transverse sections of the right-hand inferior permanent incisors of a five-year-old horse. A, forms flattened from in front to behind ; B, oval forms ; C, rounded forms ; D, triangular forms ; E, biangular forms.

D. It becomes triangular and has three (3) borders, one anterior and two lateral. The summit of the triangle looks backward. (Sections 8 to 11.)

E. Finally the table surface is flattened from side to side. (Sections 12 to 16.) This last form characterizes very old age, and lasts for the life of the animal.

Girard designated this form of the tooth as biangular. Distinctive shape in the dental table is much more regular, and gives a more reliable source of judging the age in the pincher teeth than in the intermediate, and more so in the latter than in the corner teeth. The same is true of the incisors of the upper jaw, but it is rare that we examine the table of these as an aid in judging age, except in old animals from sixteen years to twenty and over.

STRUCTURE OF THE INCISORS.

The study of the structure of the incisors of permanent dentition furnishes valuable indications for determining the age of the horse. This study is an indispensable adjunct for the complete appreciation of the descriptions which have preceded, and will allow us to understand better the peculiarities of form which we find on the table of the tooth in its changes from year to year.

If we take as a type an inferior incisor at the moment of its first appearance in the body of the jaw-bone, we find that it commences as a germ, which consists of a sack (germ-follicle) into the interior of which protrude two papillæ,—one superior, *a*, and one inferior, *b* (Fig. 11), which look toward each other.

The first, *the germ of the enamel* (*a*), is of a conical form, and is placed in what will become the dental cup.

The second, *the germ of the ivory* (*b*), depressed from in front to behind, fills the cavity which hollows out the root of the tooth.

The surface of the first is destined to the formation of the enamel of the tooth, which function it completes by the time the tooth has appeared through the gum, when the papilla disappears.

The other. which forms the ivory or dentine, remains until a more or less advanced period of the life of the animal, and furnishes the tooth with its vitality.

In the interspace between these two papillary systems is deposited the matter which constitutes the bulk of the tooth (Fig. 11, *d*). This at first consists of a very thin, conical plate hollowed out on the inside, and containing a deep depression on its free extremity. Later, the walls of the dental follicle (Fig. 11, *e*) are transformed into the alveolar periosteum. If we refer now from Fig. 11, where we find *a*, the papilla of the enamel; *b*, the papilla of the ivory; *d*, the plate consisting of the deposits of these two papillæ, —that is, the enamel and dentine juxtaposited,—and *e*, the germ sac, which is to become the alveolar periosteum. to Fig. 12, which is a longitudinal antero-posterior section of an inferior permanent pincher tooth, fully developed (somewhat enlarged), we find *E*, the enamel of the cup. surrounding this cavity. from which the papilla has entirely disappeared; *P*, the cavity of the ivory papilla very much diminished in size; *I*, the dentine occupying the major portion of the tooth. having developed on the inner surface of the thin plate which we found in Fig. 11. In Fig. 12, at *T*, which represents the table of the tooth after the borders have been worn. we find a separation of the enamel of the cup and the ivory enamel (Fig. 12, *E E*). which we find continuous in Fig. 11, *d d*.

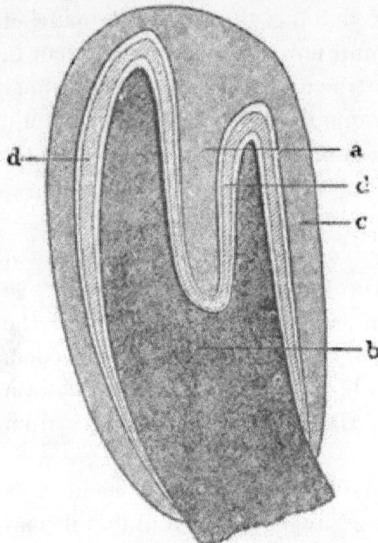

FIG. 11.

Schematic section of the dental follicle of an inferior incisor in the horse. *a*, superior papilla,—germ of the enamel; *b*, inferior papilla,—dental pulp or germ of the ivory; *e*, wall of the follicle; *d*, plate of enamel and dentine juxtaposited.

Three substances enter into the structure of the teeth, — one fundamental, the dentine or the ivory, and two covering substances, which differ from each other very much, and are known as the cement and the enamel.

1. THE CEMENT (Fig. 12, *C C*).—The cement forms the most superficial layer; it is deposited directly upon the enamel over the whole surface of the tooth, and dips into the cup, which it fills more or less completely in different subjects. Sometimes it is excessively thin, 2 to 3 mm.; at other times it is from 10 to 15 or even 20 mm. in thickness. It is generally thicker in the lower teeth than in the superior. MM. Chauveau and Arloing, in France, and Mr. Mayhew, in England, were the first to call attention to the importance of this in judging age.

It will be readily seen that the size of the cavity in the dental cup depends greatly upon the amount of the cement which is deposited in it. It is

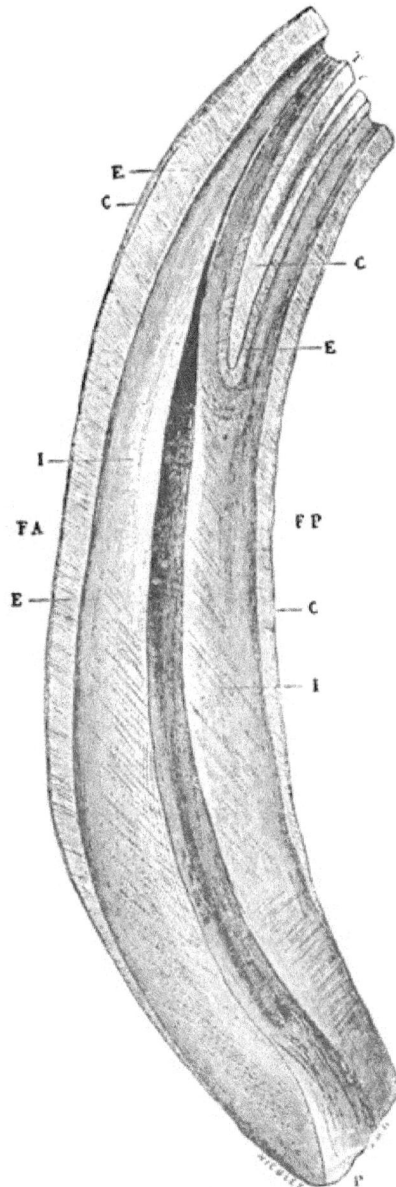

FIG. 12.

Longitudinal and antero-posterior section of permanent incisor (enlarged). F A, anterior, F P, posterior face; C, cement; E, enamel; I, ivory, or dentine; T c, external dental cavity, or cup of enamel lined with cement; P, internal dental or pulp cavity.

exceedingly rare to find excess of depth due to increased depth
of the enamel, which condition will, however, be considered in
the study of the abnormalities of the teeth.

The cement is a protecting layer which offers only a moderate resistance to the friction of food and other substances, and
disappears at an early date from the periphery of the tooth, while

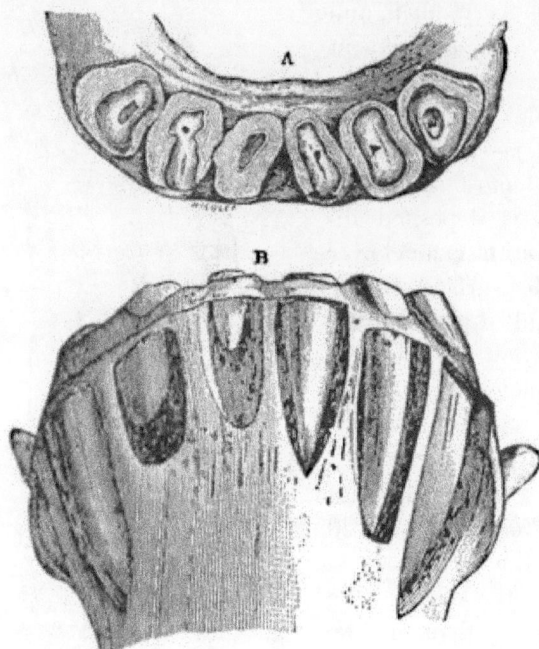

FIG. B.

Cementation of the roots of the incisors. A, dental tables; B, roots in their
alveolar cavities.

it persists in the cup of the tooth so long as the latter remains
on the table, where it forms a whitish spot, surrounded by a
band of enamel. (Fig. 10, *A*, *B*, and *C*.) The cement is shown
by microscopical examination to be a bony formation, secreted
by the alveolar periosteum (germ sack), and is not a transformed
ivory or dentine as given by Simonds.

It is found in greatest quantity at the crown or free

extremity of the tooth ; but in very old horses, when the teeth
have worn down to the roots, and the dentine, no longer protected by the enamel, offers a diminished resistance to the friction to which they are exposed, the irritation of the dormant
periosteum sets up a new secretion, and cement is thrown out
often in very great quantities to strengthen weak roots. As the
inferior incisors are the shortest, and the first to wear away, we
more frequently find the large masses of cement in the lower
jaw. We again find excessive deposits of cement around the
roots of the teeth when they have been loosened or irritated in
their cavities by the rough pressure of bits, or have been injured
by any accidental cause. This occurs most frequently in vicious
or playful horses which have the habit of biting roughly at
foreign bodies. We see in this a wise provision of nature to
succor these important organs after accident, and from the
effects of old age.

2. The Enamel (Fig. 12, *E E*; Fig. 11, outer surface of *d*).
—The enamel is the true protecting layer of the teeth. Under-
neath the cement it forms a sort of armor which covers the surface of the dentine and forms the walls of the cup. It does not
reach the cavity of the pulp. It is thicker, and covers a greater
extent on the anterior face of the tooth than it does on the posterior. (Fig. 12, *E E*.) This furnishes an important factor in
determining the age of very old horses, when the tooth either
becomes triangular or biangular. In the wall of the cup the
enamel has about the same thickness throughout, although the
difference of a transverse or an oblique section may give a
deceptive appearance of a greater thickness at one point than at
another

The enamel has a wonderful hardness. While still inclosed
in the germinal sack it is readily cut by a knife and its elements
can be dissected into Z-shaped prisms. From the moment
that it is exposed to the air it becomes so hard that it will strike
fire from flint. As it is more resisting than the dentine, it
constantly stands in relief on the surface of the table of the
tooth.

HISTOLOGY.—The enamel is of epithelial origin, formed from the superior papilla of the primitive follicle. On microscopical examination it is found to be composed of an infinity of little hexagonal prisms, intimately joined and directed obliquely to the subjacent surface. The deepest layers lie immediately over the peripheral lacunæ of the dentine.

3. THE FUNDAMENTAL SUBSTANCE, DENTINE, EBURNATED SUBSTANCE, OR THE IVORY (Fig. 12, *I*).—This constitutes the major part of the tooth. It is produced by the inferior papilla or pulp, has a deep depression in its free extremity for the dental cup, and is covered by the enamel. It forms the walls of the pulp-cavity. In the forming tooth, the dentine consists of a thin layer juxtaposited to the layer of enamel formed by the superior papilla (Fig. 11, *d*); but as the tooth develops, successive layers are deposited on the interior of the first, and the ivory gradually encroaches upon and diminishes the size of the papilla lodged in the pulp-cavity until the latter entirely disappears.

The deeper layers have a darker color than the first, as in their continual infringement on the vascular papilla they imprison a portion of its organic matter and blood-supply until they have produced a complete atrophy of their progenitor.

This discoloration of the dentine we shall find later, on the table of the tooth, under the name of the "dental star," which becomes an important factor in judging of age after eight years. Strictly speaking, there are two stars, but the posterior is very rudimentary and of no practical importance. It is occasionally, however, seen very distinctly.

In Fig. 14 we have a series of longitudinal antero-posterior sections of incisor teeth from horses of different ages, which show the gradual diminution of the pulp-cavity from the encroachment of the dentine. The dentine toward the root of the tooth is always of a darker color than that toward the crown. The ivory is less hard than the cement, but much more resisting than either the cement or the bone in which the tooth is imbedded.

HISTOLOGY.—The dentine is channeled by a multitude of *canaliculi*, which radiate from the pulp and after frequent anas-

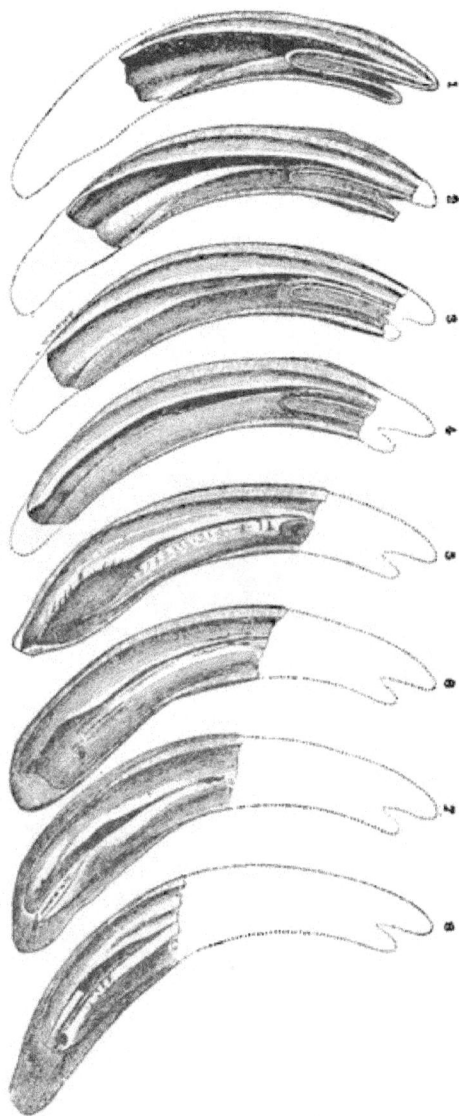

FIG. 14.

Antero-posterior and longitudinal sections of incisor teeth of horses of different ages, showing the gradual diminution of the pulp-cavity and the wearing of the free extremity. 1, 2, 3, 4, 5, 6, 7, and 8 are longitudinal antero-posterior sections of incisor teeth from horses of 3, 5, 7, 9, 13, 20, and 25 years of age, shewing the growth of the teeth at their roots; the progressive wearing of the tables; the length and obliquity at different ages; the obliquity of the pulp-cavity; the cementation of the roots.

tomoses terminate in lacunæ under the deep layer of the enamel (*interlobular spaces of Czermak*).

THE TABLE OF THE INCISORS.—The table of the incisors is

the free portion or crown of these teeth, which becomes worn by friction with the hard substances which the animal takes as food, and by the constant contact with the teeth of the opposite jaw. After the high anterior and lower posterior borders of the virgin incisors have been worn away the table is established. At first this consists, on each tooth, of an oblong plate, wider from side to side, which is surrounded by the layer of cement directly covering the peripheral layer of enamel; inside of this is found the zone of yellowish and softer dentine, usually somewhat depressed on account of its lesser resistance to friction; in the

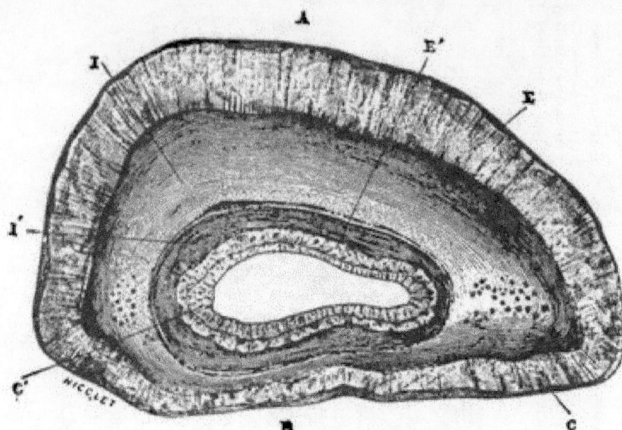

Fig. 15.

Transverse section of an incisor tooth. Inferior, right side. A, anterior; B, posterior; E, peripheral enamel; E', central enamel; C, peripheral cement; C', central cement; I, ivory, or dentine. (Enlarged.)

middle, the central enamel or border of the cup surrounds the variable quantity of cement which may be deposited in it. (Fig. 10; Fig. 12, A; Fig. 15.)

As the table gradually encroaches upon the wedge-shaped tooth (Fig. 16), it becomes proportionately narrower in its transverse diameter, and as it becomes oval in shape the cup is found nearer the posterior border of the tooth instead of in the centre. The dental star, or dark-colored dentine which takes the place of the papilla, now appears between the cup and the anterior border of the tooth (Fig. 12, P; Fig. 10, B.). A still further

destruction from use brings the table to the middle of the
original tooth; when it assumes the round form, the cup of
central enamel has disappeared, and the dental star is now midway between the anterior and posterior borders. (Fig. 10, *C*.)

Still further use brings the triangular form, and finally the
biangular; in these two forms the dental star gradually becomes
larger, and, at times, when the papilla has not been entirely
replaced by the dentine, a cavity is found which, by the un-
informed, has been mistaken for the cup (Fig. 10, *C* and *D*), but
should not be, as there is no surrounding enamel. Figure 16
shows the forms at the various parts of the original tooth.

THE DIRECTION OF THE INCISORS.—
The direction of the incisors, or the position which they hold in regard to the jaw,
is to be studied first in profile, considering
the relative angle which the incisors of
the upper and lower jaws have to each
other; and second, in face, considering the
position which they hold in regard to the
median line.

DIRECTION OF THE PLANE OF CON-
TACT OF THE TWO JAWS (Fig. 17).—In a
young horse the incisors meet and form
an arch which, if viewed in profile, represents the half of a circle; so that a
tangent drawn from the point of contact
of the two jaws is perpendicular to their

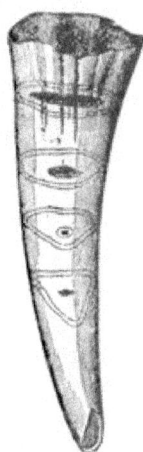

FIG. 16.
Scheme of sections of wedge-
shaped incisor.

tables. But as the progressive wearing of the table brings
it nearer to the roots of the teeth, the half-circle changes to
the form of an ogive, which becomes more and more acute
as the surface of contact, which is displaced above and below
parallel to itself, extends gradually from the primitive diameter.
Consequently the tangents, *a a'*, *b b'*, *c c'*, drawn from the new
points (*a*, *b*, *c*) of contact of the arches, are no longer perpendicular to this point, but tend to become parallel.

As the angle of incidence of the incisors increases in

obliquity with age, and the horizontality of their arches indicates approximately the degree of the alteration,—excepting,
of course, certain abnormal changes to be studied later,—we
have in the profile of the jaw a valuable factor by which to
judge of the age of the horse.

DIRECTION IN REGARD TO THE MEDIAN LINE (Fig. 18).—
In the young horse, the crowns of the six incisor teeth widened
from side to side, while their roots are flattened in their trans-

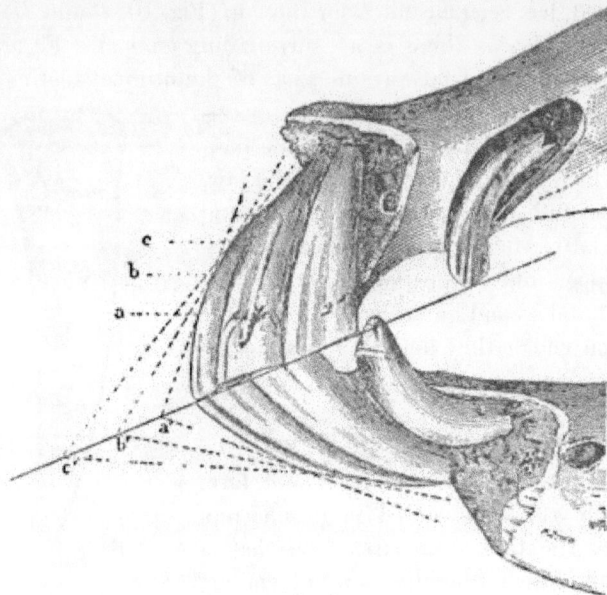

FIG. 17.

Arch of the incisors, uncovered to show their successive planes of contact.

verse diameter, cause the teeth to take a position on the end of the
jaw like the ribs of an open fan. They diverge from the alveolar cavities, in which they are imbedded in the bone, toward the
circumference formed by their crowns. But, as age advances,
the crowns become worn off, and the teeth are pushed farther
and farther out of their cavities. The roots, which were at first
almost in contact with each other, gradually separate and widen,
while the circumference constantly diminishes as the teeth

become worn down, until they assume a parallel position, and
finally converge, instead of diverging, at their free extremities.
The intermediate teeth become separated from the pinchers on
the one side, and the corner teeth
become separated from them on
the other, until a distinct space
is visible between them, which is
filled by a pale gum.

En résumé in regard to the
general direction of the incisors:—
1. The incidence of the
arches acquires a greater obliquity
with age. 2. Their incurvation
and their transverse diameter diminish. 3. The teeth, at first diverging from their roots, become
parallel and finally converge
toward their free extremities.

THE TUSHES, TUSKS, CANINE TEETH.

The tushes are four in number in the adult horse; they are
rudimentary or absent in the
mare. In certain sections of the
country superstition has attributed
sterility to mares provided with
them, which is, however, absolutely unfounded.

The superior tusks are placed
one on each side of the upper interdental spaces, at the point of
union of the superior maxillary

Fig. 18.
Incisors of the two jaws (roots uncovered), seen by their anterior face to
show their direction in regard to the
median line.

and intermaxillary bones. The lower ones are placed on either
side of the maxilla, always closer to the incisor teeth than those
of the upper jaw, so that they are invariably in front of the latter.

The canine teeth are curved in their long axis with the

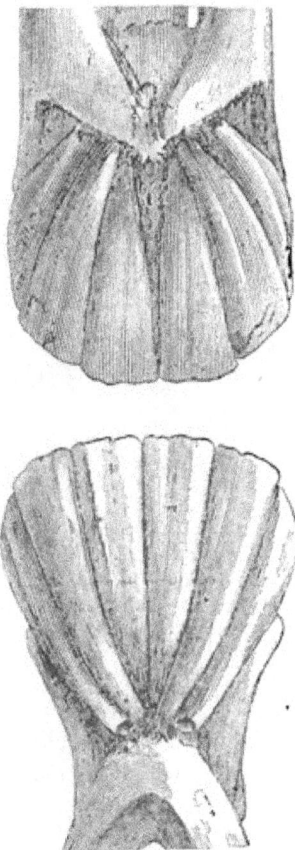

concavity backward, and they project slightly outward. The external face is covered with small, longitudinal, parallel striations. The internal face has a conical eminence pointing toward the free extremity of the tooth and separated from its borders by deep gutters. The tusks have a pulp-cavity relatively much larger than that of the incisors, which, however, also becomes filled with a discolored dentine as the tooth advances in age. As the tusks become used they show in their centre a dental star, which sometimes is very dark in color. While the tusks

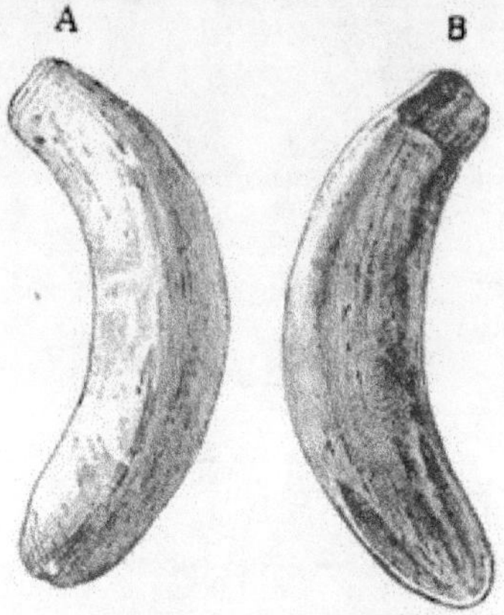

FIG. 19.
Upper right-hand canine tooth of a horse.
A, external face; B, internal face.

FIG. 20.
Longitudinal antero-posterior section of a horse's canine tooth. C S, superior canine; C I, inferior canine.

may be subject to much use from the friction of food, the tongue, lips, and the bit, they are always prominent, even in very old horses. Goubaux and Barrier have seen one case in which the tusks of the upper and lower jaw were in contact.

STRUCTURE.—The tusks are composed of ivory or dentine, inclosing a pulp-cavity and surrounded with enamel.

There are no temporary tusks in the horse, although small, rudimentary spicula sometimes are found before the permanent teeth appear.

MOLARS.

PREMOLARS—WOLF-TEETH.

THESE teeth, which were first described by Daubenton, are not found in all horses. They occur much more frequently in the upper jaw than in the lower one, and are rarely found in both in the same animal. Girard says that they usually appear about the tenth month, and drop out at the time of the eruption of the permanent molars. The alveoli are usually directly in front of those of the first molars. They are, however, frequently a half-inch or more in front of the others, and often remain until the animal has reached an advanced age. Their presence renders the dental formula slightly variable (forty-four for the horse and forty for the mare). The wolf-teeth are elongated and slightly curved on their long axis; they resemble a good deal the incisive teeth of the carnivora, from which, perhaps, comes their common name. Their roots are unicuspid. Girard describes supplementary post-molars, which, however, are not recorded by other authors, and were probably anomalous teeth.

MOLARS.

The molars fill in the sides of the dental arch; like the incisors, they appear in two groups,—the temporary and permanent teeth.

MOLARS OF FIRST DENTITION.

The *deciduous, temporary, milk,* or *molars of first dentition,* are twelve in number,—six in each jaw, three on each side. By some curious error, the father of veterinary anatomy, Carlo Ruini, 1598, and his immediate followers, considered that there were but eight temporary molars.

These teeth are strong, short, and have the general form of quadrangular prisms, except the first in each row, which is triangular.

The anterior and posterior faces are smooth; in the upper jaw the first tooth has three longitudinal canals on its external

3

(33)

face, while the second and third teeth have two; the internal faces are irregular, slightly concave in their long axis, and show canals very much less marked than those on the outside.

The inferior teeth on their external face have a single gutter, which is deepest in the first tooth and most shallow in the last; on their internal face they are irregularly grooved. All of the milk-molars have a constriction or neck separating their crowns

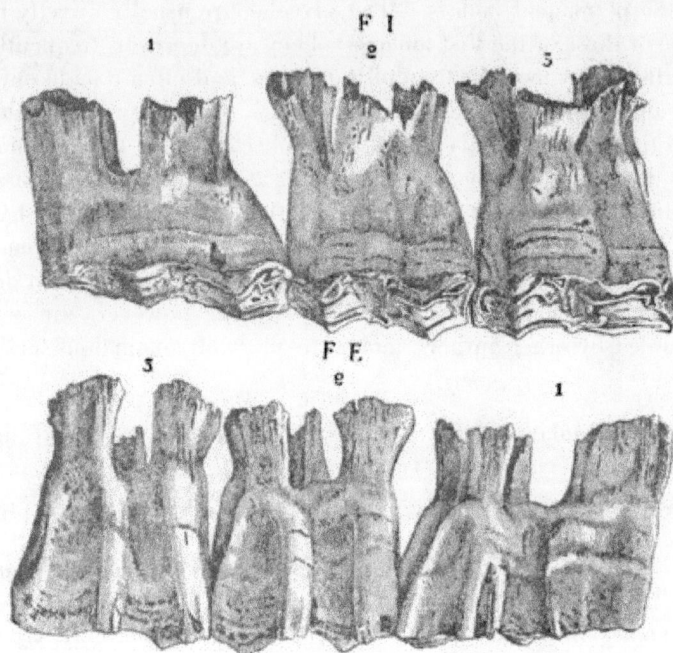

FIG. 21.
The three superior temporary molars (right). F I, internal face; F E, external face.

from their roots. They have each two roots, one anterior and one posterior. These are strong, convex outside and concave inside; each is hollowed by an opening which reaches into the tooth itself. Just before being forced out by the permanent teeth the roots are sometimes divided into little eminences by the pressure of the irregularities on the crowns of the latter. The free extremities of the virgin milk-molars are irregular and covered with ridges and cavities, but the external border of the superior teeth and the internal border of the inferior teeth are always longer than the opposite border, so that they present two oblique planes, which become more marked as the teeth wear down. As the animal becomes older the molars wear down till they are reduced (Figs. 23 and 24) to little thin plates.

F I

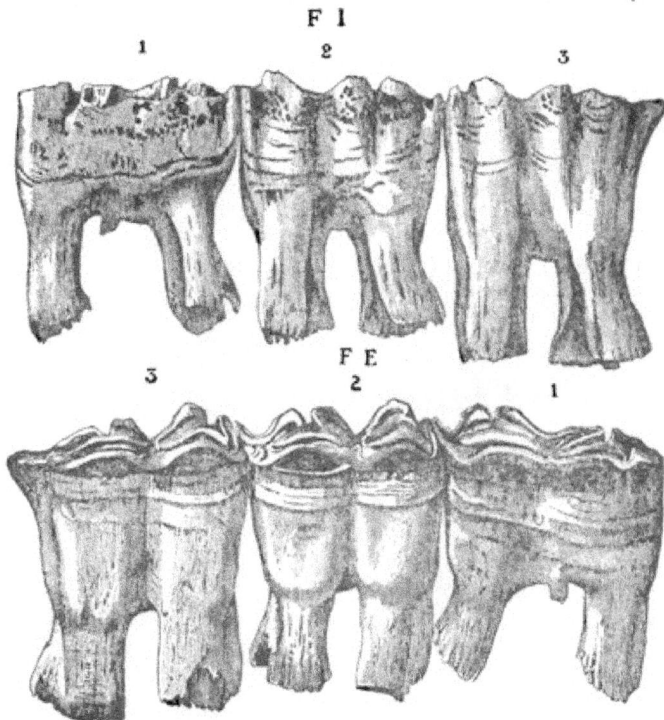

FIG. 22.

The three inferior temporary molars (right side). F I, internal face; F E, external face.

which fit close to, or cap, as it were, the crowns of the replacing molars, by which they are finally forced out from the jaw.

MOLARS OF SECOND DENTITION—PERMANENT MOLARS.

These are twenty-four in number,—twelve in each jaw, six on each side. They are designated numerically from in front to behind, as the first, second, etc. The three first, which replace the temporary molars, are known as *pre-molars*, and the last three as the *post-molars*. Each row of six molars forms a branch of the dental arch, which in the upper jaw is slightly convex to the outer side, while in the lower jaw it is perfectly straight, with the anterior extremities inclined toward the other, so that they form a sort of **V**, and are overlapped by those of the upper jaw.

FIG. 3t.

A, transverse section of the lower jaw, showing the third temporary molar and its relation to the permanent molar which replaces it. B, third inferior permanent molar, capped by the worn temporary molar.

The permanent molars have the shape of quadrangular prisms flattened from side to side, except the first and sixth, which are triangular.

SUPERIOR MOLARS.—The posterior faces are almost smooth, except in the sixth molar, where the face is replaced by a blunt edge; the anterior faces are also smooth, except that of the first molar, which is replaced by a sharp border.

The external faces have two longitudinal gutters separated
by a ridge; the first molar has three gutters with two ridges;
the internal faces have two shallow gutters on the first molar
and one on each of the others. which is closer to the posterior
border of the teeth in the last ones.   There is no neck or dis-

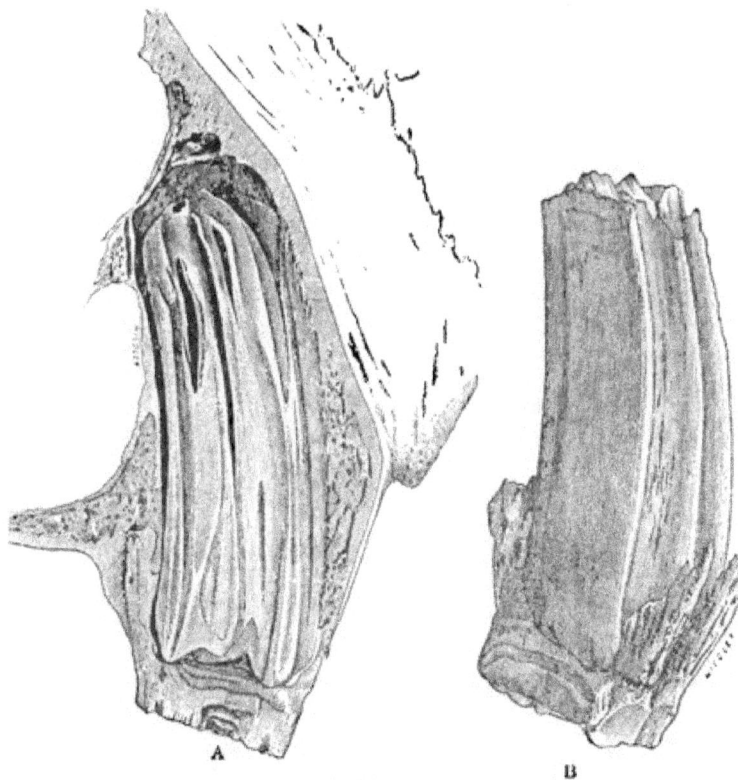

FIG. 24.

A. transverse section of the upper jaw, showing the left third temporary
molar and its relation to the permanent molar which replaces it. B. third permanent molar, capped by the worn temporary molar.

tinctive line (Fig. 24) separating the crown or free extremity
from the roots or imbedded extremity of the permanent molars.
The free extremity is triangular in the first and sixth teeth, and
quadrangular in the others.   In the virgin tooth it has an irregular surface resembling an Old English **B**. with loops looking

FIG. 25.

Right-hand superior molars of a horse 6 to 7 years old. First molar at top of figure.

FIG. 26.

Left-hand inferior molars of a horse 6 to 7 years old. First molar at top of figure.

toward the inside and surrounding cavities, which are more
or less filled with cement; to the anterior loop is attached a
small, secondary loop. The external border is always longer
than the internal. As the tooth wears and the table is established, the cavities disappear. At first the imbedded portion is
hollowed by cavities, which reach the body of the tooth itself
and contain the papilla, or pulp, formed principally of blood-
vessels and nerves; later, these resolve into distinct roots, three
each for the first and sixth molars, and four each for the others.
According to Girard, these become distinct at five years of age.

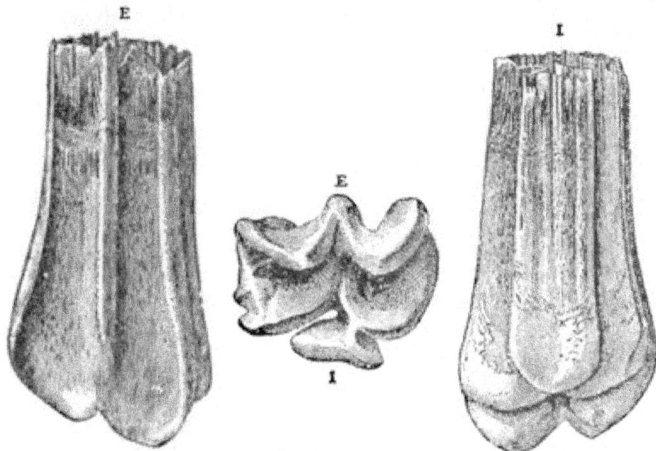

FIG. 27.

Superior permanent molar (right side).—virgin tooth. E, external face;
I, internal face; E I, surface of contact.

(Fig 27.) If the head is placed horizontally the first molar is
found imbedded in a vertical position, while the others incline
somewhat from below to above and from in front to behind.

The superior molars are implanted in alveolar cavities, prismatic like the teeth, and separated from each other by bony
septa, which are thin at their free borders and thicker above.
The bottoms of the three last project into the maxillary sinuses,
while those of the three first are in the superior maxillary bone
itself. In very old horses the sinus may reach the root of the
third or even second tooth.

INFERIOR MOLARS.—The anterior and posterior faces are flat, except the posterior of the sixth tooth and the anterior of the first tooth, which are replaced by sharp edges. The external faces have one longitudinal gutter in the first five teeth and two in the sixth tooth. The internal faces have three gutters in the first and sixth teeth and a variable number in the others. The free extremity is triangular in the first and sixth molars and quadrilateral in the others; but these are narrower from side to side than from front to behind. They are longer on their internal border than on the external, and have the Old English ℬ turned with the loops outward. The imbedded portion is bicuspid in the first five teeth and unicuspid in the sixth. The divisions of the roots diverge; each contains a cavity for the pulp.

DEVELOPMENT AND STRUCTURE OF THE MOLARS.

In the superior molars the papilla of the enamel in the

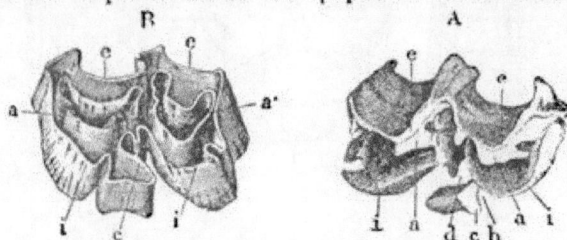

FIG. 28.

A right superior molar taken from the dental follicle   A, view from free end ; B, view from the imbedded portion ; a, anterior infundibula ; a', posterior infundibula ; e, external border of the infundibula ; i, internal border of the infundibula ; c, accessory column of the anterior infundibula.

germinal sack is double and penetrates into the tooth, forming, practically, two infundibula, one anterior and one posterior, instead of one cup, as in the incisor teeth. The papilla of the pulp makes five diverticula, which almost appear to be as many separate cavities.

In the inferior molars there are two infundibula, one occupying the middle portion of the tooth and one toward the anterior portion, which allows us to recognize a left-hand inferior molar from its right-hand homologue. The arrangement of the infundibula of the papillæ of the root is practically the same as in the superior molars.

The molars. like the incisors, are composed of a fundamental substance and two covering layers.

FIG. 29.

Grinding table of a superior molar (left).

Table of an inferior molar (right).

A. *The Enamel* forms the bulk of the tooth. It covers the four faces and is reflected into the infundibula. On the table

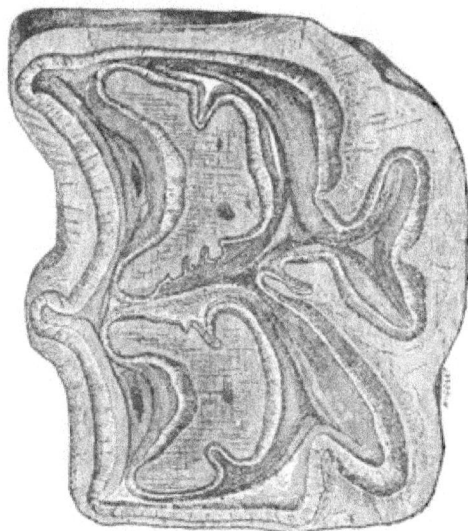

FIG. 30.

Transverse section of superior molar (left). Enlarged.

of a molar which has been worn down it forms the bright lines which form the Old English 𝕭.

B. *The Ivory or Dentine* is deposited on the internal face of the enamel. and finally fills up the diverticula of the pulp-cavity. At first it is entirely protected by the enamel, but as the table is formed it becomes exposed, and, as it is softer than the other substance, becomes worn more rapidly, and renders the wearing surface uneven and better adapted to grinding.

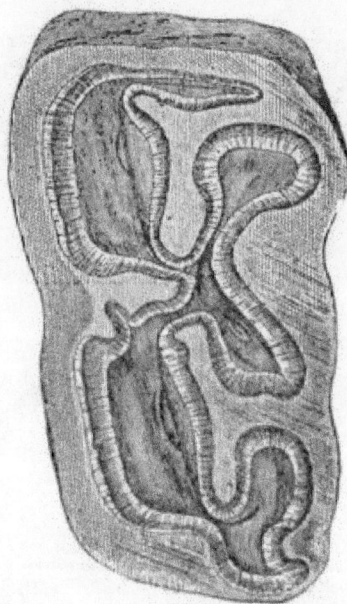

Fig. 31.
Transverse section of an inferior molar (right). Enlarged.

C. *The Cement* is very abundant on the molars. It covers the enamel, fills the infundibula of the enamel, and, in very old mouths, is often formed in excess, and furnishes a new wearing surface to replace the teeth themselves, which have been destroyed by use.

ERUPTION OF THE TEETH.

THE eruption of the deciduous incisors does not cause any marked constitutional phenomena, although it may cause some slight loss of appetite,—due, however, rather to the local pain in chewing than to any general disturbance. With the appearance of the permanent teeth the head increases in size and becomes more full, which is due principally to the space demanded for the large molars of second dentition.

The appearance of the permanent teeth is a period of general irritation of the animal economy. The first attack of periodic ophthalmia frequently appears at this time; colts seem more susceptible to contract strangles and other of the contagious diseases.

Sometimes the temporary incisors drop out before the appearance of the permanent teeth; in this case they leave a small ulcer in the gum, which is followed by a swollen, red, sore point, through which finally the replacing tooth appears by its internal corner, which is followed by the anterior border, set not in its future proper position, but obliquely to the axis of the jaw. When the temporary teeth have been pulled to hasten the age, as is a growing custom in some of our neighboring markets, the obliquity remains, and when found in an apparent five- or six-year-old mouth should always be regarded with suspicion. At other times the milk-teeth remain, and the permanent ones protrude on their posterior border. Convulsions may occur by excessive pressure and irritation in these cases. The incisors of the upper jaw usually appear before those of the maxilla. The pincher, intermediate, and corner teeth appear in succession by pairs.

Causes which can Hasten or Retard the Eruption of the Incisors of the Second Dentition.—All ordinary horses are

(43)

supposed to be born on the 1st of May, while thoroughbreds are supposed to be born on January 1st, and each dates its age accordingly. We may, therefore, have horses which officially have the same age vary from each other in their real age by some months.

Races which develop slowly, a debilitated temperament, and an impoverished nourishment, all tend to retard the eruption of the teeth, while, on the contrary, precocious races, strong feeding, etc., hasten the appearance of the full mouth. Gestation in the young mare retards the eruption of her teeth, especially of the corner teeth, which mark the five years. In cold, damp climates the teeth are several months later in appearing than they are in warmer and dryer climates. The thoroughbred and all improved horses develop their full mouths earlier than their less well-bred relations. Exceptions may occur in both extremes. A rising three-year-old has been known to have all of its permanent incisors, and more frequently a six- or seven-year-old is seen with its corner milk-teeth.

ERUPTION OF THE TUSKS.

The appearance of the tusks is so variable that it is of little value as an indication of age. They are usually absent in the mare, and in the horse or gelding may be out at three years, or may not pierce the gums until the animal has had all of its other permanent teeth. The *lower* tusks frequently appear almost a year in advance of the upper ones. They usually appear at about the same time as the corner incisors.

ERUPTION OF THE MOLARS.

As the young horse becomes older the deciduous molars become worn down until they only exist as thin plates which cap the crowns of the first three molars of permanent dentition and are but loosely held in their alveola. They become broken, and the sharp points wound the neighboring cheek, causing difficulty of mastication, which can frequently be relieved by removal of the loosened teeth.

The two first molars of first dentition are through the

gums at birth. or within a few days of it. The third appears at the end of the first month. The authorities are not in accord as to the appearance of the permanent molars, as will be seen in the following table:—

DESIGNA- TION OF TOOTH. PERMA- NENT MOLARS.	EPOCH OF ERUPTION.		DESIGNATION OF TOOTH.		EPOCH OF ERUPTION.
	GIRARD.	MAYHEW.	INFERIOR MOLARS.	SUPERIOR MOLARS.	SECELLIER.
4th	10 months.	12 months.	4th	4th	10 and 12 mos.
5th	20 months.	18 and 24 mos	5th	5th	20 and 24 mos.
1st	30 and 32 mos.	. . .	1st and 2d	1st	30 and 36 mos.
1st and 2d	. .	36 months.	6th	6th	32 and 36 mos
2d and 3d	36 months.	.	3d	2d	40 and 42 mos.
6th	4 and 6 years.	. . .	. . . . .	3d	44 and 48 mos.
6th and 3d	. . . . . .	60 months.			

According to Secellier, the inferior milk-molars fall constantly before those of the upper jaw, while the eruption of the permanent molars of both jaws takes place at the same time.

ACCORDING to Buffon, the duration of the life of the horse is, as in all other species of animals, proportionate to the duration of its growth. Man, who is fourteen years in growth, lives six or seven times that length of time; that is to say, he may live to 90 or 100. The horse, which requires about four years to attain its growth, may live six or seven times that length of time; that is, to twenty-five or thirty years. Examples which are contrary to this rule are so rare that they need hardly be regarded as exceptions of consequence; as the more common draft-horses acquire their growth in less time than better-bred horses, they also live a shorter time,—fifteen or sixteen years.

According to Bourgelat, "the age of the horse can be estimated at eighteen to twenty years. The number of those which pass this age is excessively small."

Aristotle noted that horses which are nourished in stables live for a shorter time than those which are raised in herds; the conditions of slavery and domesticity diminish their powers of resistance to the wear and tear of life.

Athenæus and Pliny claim to have known the horse to attain the age of sixty-five or even seventy years. Augustus Nipheus speaks of a horse belonging to Ferdinand I which was over seventy, but these observations are only exceptions similar to those which we sometimes find in the human species.

Hartman and Buffon both note that the life of mares is generally longer than that of horses. This observation was already made by Aristotle (Hist. Animal., lib. v), and corresponds to the same rule in the human race, where women generally live to a greater age than men. Hartman claims that it is "an undoubtable sign that a breeding horse is well bred and in good health when it is slow in development. Those which do not attain their complete growth until six or seven years, excepting in cases of accident, are useful for twenty years or more,

(46)

and may still be sound and healthy at thirty years. The other
ones, on the contrary,—those which attain their growth in four
years,—rarely pass the age of twenty."

The great, heavy draft-horses, which attain their growth in
even less time, are already aged at ten or twelve years. Examples of horses of thirty, thirty-six, and forty years of age would
not be so rare among our animals if the tyranny and hard usage
imposed upon them by men did not aid greatly to shorten their
lives. Ordinarily, as soon as a horse has seen its best days, it is
sold from the stable in which it has been kept, to be replaced by
more useful animals, and it goes rapidly into the hands of the
hackman and the huckster, doing harder work with less nourishment, until it becomes completely used up, and depreciates to
the value of the knacker's price.

Among the principal causes which modify the longevity of
the horse may be counted the length of time of development,
the size of the horse, the work to which the animal is put, and
the care which it receives.

We find certain races and certain individuals which are precocious in their development, and other races and subdivisions
of families in which longevity is hereditary.

M. Bouley says: "There are tardy races and precocious
races. In the last the precocity is the result of the combined
influence of heredity and alimentation, in which the organic
formation acts in a precipitous way, as it were, in those individuals which compose it, and we find a hasty achievement of
development; from which it results that the duration of their
first part of life is so much shortened, and, as a fatal consequence, also that of their whole life; for this more rapid development, impressed on their organisms, has no other termina-
tion, from the industrial point of view, which produces it, than
to hasten the moment of death."

The small races of horses and small individuals last usually
for a longer time than larger ones. We can make no satisfactory explanation of this, unless it is that in the larger animals
the wear and tear is greater from the very fact of the animal

having so much more to take care of in itself. The character of work to which the horse is put is an important factor in shortening the life,—due to the fatigue entailed and the using up of the powers of the organs. We find that a horse with a quiet and docile disposition will perform heavier and rougher work, with less deterioration in its animal economy, than one alongside of it of a nervous and excitable disposition. The more calm and quiet life of horses in the country is less prejudicial to their existence than the bustle and confusion of the great industrial centres and crowded streets of our large cities.

Again, the necessities of rapid and variable work and irregular hours of feeding in great cities predispose to internal diseases, while the greater loads and rough pavements lead to strains of tendons, wrenches of bones, bruised feet, and other injuries which soon render the animals unserviceable, or condemn them to return to the softer ground, lighter work, and more regular life of the country, with, however, a broken-down constitution. The care which the horse receives from its keeper—which, of course, includes the regulation of work and feed—is an all-important matter. A kind, judicious owner will work horses for years and keep them sound and in good condition, doing a vast amount of work; while in the next stable, under the guidance of a careless or brutal owner, equally good horses will perform less work and become worthless in a short time.

The word "used" applied to horses has become synonymous with the word "aged" applied to man, and indicates the time when the animal has become prematurely old.

Examples are not rare in which the horse has attained the age of thirty, thirty-six, forty, and even more years, and has been in perfect health and capable of moderate useful service.

PRINCIPLES OF EXAMINATION FOR DETERMINING THE AGE OF THE HORSE.

To any one accustomed to horses it is an easy matter to distinguish at sight the very young from the adult horse, and the middle-aged from the very old animal. In very old animals white hairs commence to show in the neighborhood of the temples and around the eyes and the nostrils, if the color of the animals is dark; gray and roan horses become very much lighter in color, and even at times nearly white; the inferior extremity of the head becomes pointed, and the sides of the face become depressed; there is an evident change in the line of the back and loins, which become depressed (sway-backed). The *aplomb* of the legs and the blemishes found on them are common witnesses of work which has lasted for a number of years.

In addition to the examination of the teeth, which is frequently considered the only point that need be looked at to determine age, there are many other points which are of very great value.

We have seen that the molar teeth of the very young animal are deeply imbedded in the alveolar cavities of the jaw-bone, and that when the animal becomes older the teeth are gradually pushed out from these cavities as the free extremities of the tooth become worn away. The molars are pushed out and the jaws become thinner by absorption of the bone, which is no longer needed as a bed for holding the teeth, which are diminished in size. We find, then, that, as the horse becomes gradually older, the branches of the jaw-bone, which in the young animal were thick and rounded, become gradually thinner and sharp on their border, and form a very accurate indication of the approximate age of the animal.

In an old Arab book of agriculture, by Ibn-el-Awamm, written about the twelfth century, it is stated that "one of the signs by which an old horse can be recognized is to pinch between the thumb and index finger the skin of the forehead and draw it out, and then to let go of it quickly. If the skin returns promptly to its place and becomes perfectly smooth, the animal will make a good horse," etc.

In Aristotle we find the same statement prescribed as an indication of age: "If the skin returns promptly to its place and leaves no wrinkle, the animal is young; if the skin remains wrinkled, it is old."

An old and neglected, but useful, means of approximating

4

the age of old horses is by means of the "knots" in the tail.
These knots are little, prominent eminences on either side of the
base of the tail, formed by the transverse process of the coccygeal
bones. The processes can be felt in young horses, and become
especially prominent after the emaciation of a severe illness, but
in this case they are rounded and are apparently continuous with
the other tissues, while in old horses, due to absorption in the
bone itself, they become more distinct, and seem to stand out
in the muscles and softer structures of the tail.

The knots are felt distinctly at the base of the tail when
the horse has attained the age of thirteen years. In two years
later, when they have become more prominent, they have behind them a distinct little depression two or three lines in width.
At sixteen a second pair of knots are found, which, like the first,
in about two years, have behind them a distinct depression, and
so on, every three years, a new pair of knots furnish an approximately accurate indication of the age of the animal.

The teeth, however, are by far the most important witnesses of the age of the horse. The examination of these,
which includes the incisors, the tusks, and the molars, shows
(*a*) if the incisive arch is composed of teeth of the first or second
dentition, or of both ; (*b*) if there is a normal number of them ;
(*c*) the situation, length, breadth, and the angle which they
have in the maxillary bones ; (*d*) if the tables of the teeth touch
each other, or if one overreaches the other, which somewhat
modifies the leveling of the table; (*e*) the color and substances
on the face of the teeth; (*f*) if they have been subjected to
fraudulent handling.

In the incisors we study the form and details of the table,
the direction and length of the tooth, and the condition of the
corner tooth. In the tusks we note the amount of use to which
the teeth have been subjected, their direction and their length.
In the molars we look for the number and the dentition to
which they belong, the condition of their grinding surface, their
length and direction and the integrity of their substance, and
that of the gums which surround them.

CHARACTERS FURNISHED BY THE TEETH.

The reader may, perhaps, have wondered why we should have gone into such minute detail in the anatomical description of the teeth, but an accurate knowledge of all these details is absolutely necessary, when we take into consideration the slow, insensible, and variable wearing, producing often only trivial changes, which are, however, important guides.

Washed incessantly, rubbed, used, and broken as the teeth are by the action of the saliva, the lips, the cheeks, the tongue, and the food which the animal takes, we have to take into consideration every detail of their structure, their position, the breed and general condition of the animal, and the surroundings in which it has been raised. A thoroughbred with dense bones and hard teeth will wear the latter away much more slowly than a coarse-boned, lymphatic, common horse with softer substances in the teeth. The character of food to which a colt has been accustomed will stimulate or diminish the functional activity of the tooth, and, while hard substances would naturally wear a tooth faster than softer food, yet the animal which has been raised on the former will often have harder teeth than one which has not had to use them so severely.

Any one who has only been accustomed to examine the mouths of horses of one section of country will find that he must extend his ideas and adapt himself to a new condition of things, when called upon to judge of the age of horses from another region.

If a horse's mouth presents exactly the characters which indicate a certain number of years of growth, we say that it " *is* — years;" if it has not quite attained the age, it is described as " *rising* — years;" if it has passed the period and has not yet attained the markings of another year, it is counted as " — years *off*."

The natural division of the two periods of age, as indicated by the temporary and the permanent teeth, is subdivided as follows:—

1. The period of eruption of the incisors of first dentition.

2. The leveling of these teeth and their progressive use.

3. The period of the falling out of the deciduous teeth and the appearance of the permanent ones.

4. The leveling of these latter.

5. The successive forms which their tables present as the teeth become worn away.

FIRST PERIOD.

At Birth.—At the birth of the foal the incisors have not yet pierced the gums. The anterior border of the pincher and intermediate teeth can be seen under the mucous membrane, which is rendered paler than the surrounding tissues by their pressure. (Figs. 32, 33, 34.)

About One Week.—The pincher teeth have generally appeared in from six to eight days, the upper teeth preceding the lower by twenty-four or forty-eight hours. At this age the teeth are of little importance; for the general aspect of the animal, its manner of walking, which is still unsteady, and the condition of its hairs show that it is but a few days old. (Figs. 35, 36, 37.)

About One Month.—The intermediate teeth appear between thirty and forty days, just as the interior border of the pincher teeth commence to be worn. (Figs. 38, 39, 40.)

About Three Months.—There are now four teeth in the upper and four teeth in the lower jaw; the pinchers have commenced to wear on their posterior borders; they are entirely free from the gum. (Figs. 41, 42, 43.)

About Four Months.—The incisive arch has become wider; the inferior intermediate teeth are free from the gum; their anterior border has commenced to be worn away toward their inner edge as it comes in contact with the corresponding superior teeth. (Figs. 44, 45, 46.)

About Five Months.—The pincher teeth have pushed through the gums to the line of the neck of the teeth; the intermediate teeth are worn on their anterior borders; the mucous membrane is often sensitive along the posterior border of these teeth on account of the corner teeth, which have commenced to push out. (Figs. 47, 48, 49.)

(53)

About Six Months.—The intermediate teeth have pushed farther out; their posterior borders have come in contact with each other. In the region of the corner teeth the mucous membrane is puffy and congested; sometimes even at this age the anterior border of the corner teeth shows under the soft tissues.

About Eight to Ten Months.—The anterior borders of the corner teeth are seen through the mucous membrane. The intermediate teeth are entirely through the gums to the level of their neck. The inferior incisive arch forms a regular half-circle. (Figs. 50, 51, 52.) It is not important to be more precise at this age, as various causes influence and produce slight variations in the eruption of the teeth and in their leveling. Some animals are strong and vigorous, while others are weak and feeble; some have been well fed, while others have been nourished badly; and, again, we find individual peculiarities of precocity and tardiness in the eruption of the teeth as in other evidences of development. At the outset, the foal only uses the milk of its mother: at this time there is little friction and using of the teeth, except from their simple position and contact with each other. As the foal gets older and commences to use fibrous and resisting food the incisors wear away more rapidly; there is always the most use in the pincher teeth. Moreover, during this first period of the life of the animal, while it is still with its mother, other conditions allow us to judge with sufficient accuracy as to the age of the animal, which is not ready to be sold or removed until it has been weaned.

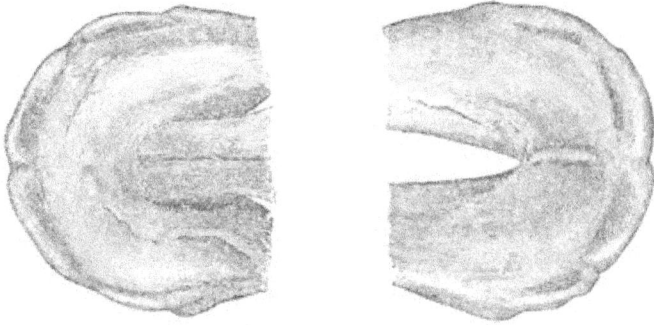

FIG. 32.

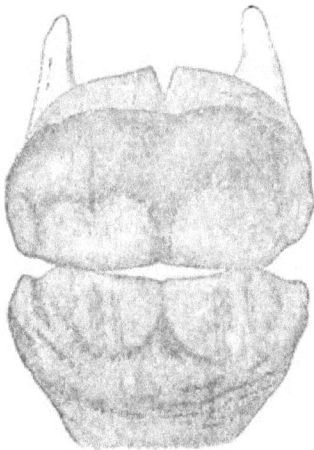

FIG. 33.

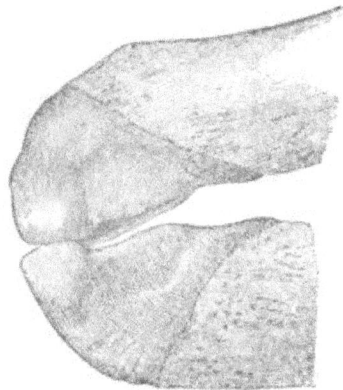

FIG. 34.

At Birth (Figs. 32, 33, 34).—The incisors are not yet out. The mucous membrane still covers the teeth, which are about to appear. In front are seen under the gums the two pinchers, above and below. In profile are seen the intermediate teeth, less developed than the pincher teeth. The jaws are rounded. The dental table shows on the side a little ridge formed by the anterior border of the pinchers, with a less distinct elevation for the intermediate teeth on the side; the edge of the teeth near the symphysis of the jaw is higher than the outer edge, and comes through first.

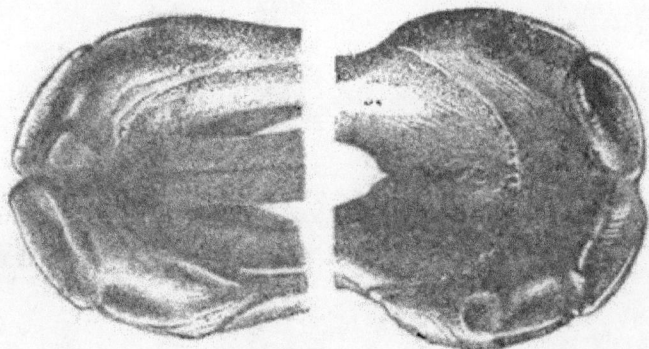

FIG. 35.

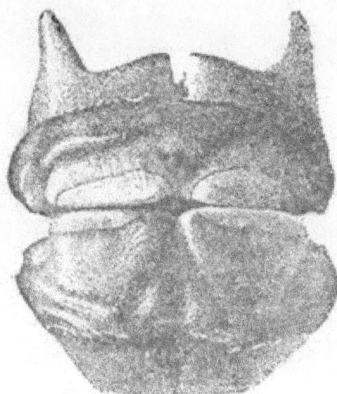

FIG. 36.

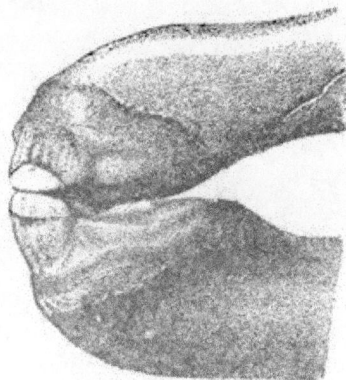

FIG. 37.

One Week (Figs. 35, 36, 37).—The pincher teeth are through the gums. In face, the anterior borders are seen. In profile, the gums are thinned on the intermediate teeth. The tables show the anterior borders of the pinchers through and depressions over the posterior borders.

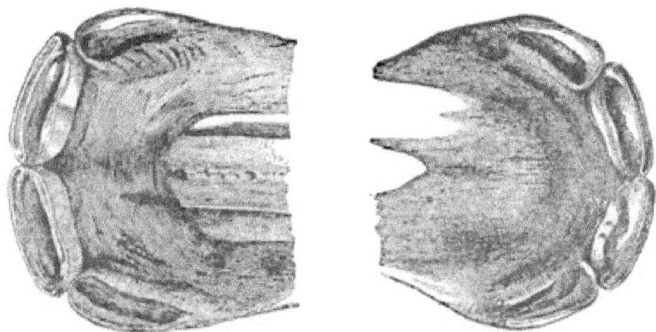

FIG. 38.

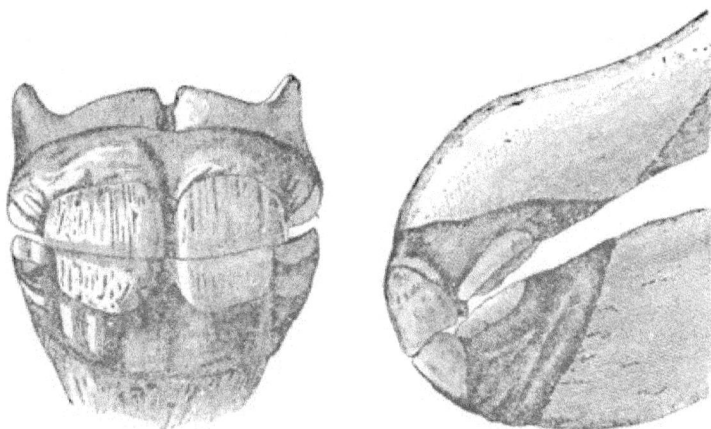

FIG. 39.   FIG. 40.

One Month (Figs. 38, 39, 40).—In front the pincher teeth, which appeared during the first week, are in contact with each other; their anterior face is striated with little gutters. On the side, the anterior border of the intermediate teeth is seen. In profile, the jaws are seen thicker; the gums still cover part of the anterior face of the intermediate teeth. The dental tables are uncovered in the pinchers. There is a slight use of both borders of the teeth above; below, the anterior border only has been worn. The membrane still covers the posterior border of the intermediate teeth and a portion of their dental cup.

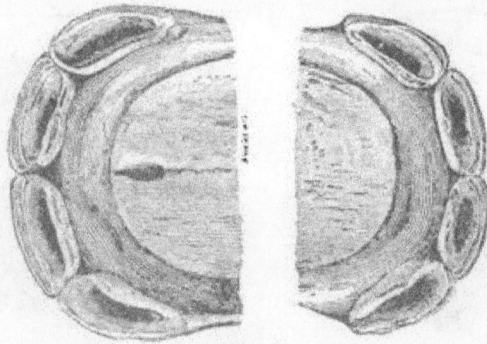

FIG. 41.

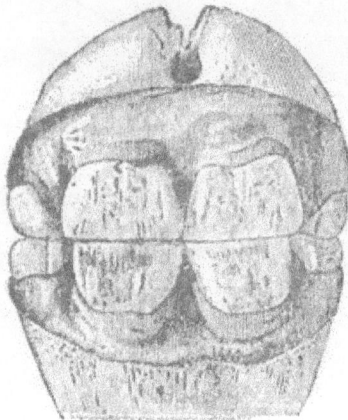

FIG. 42.

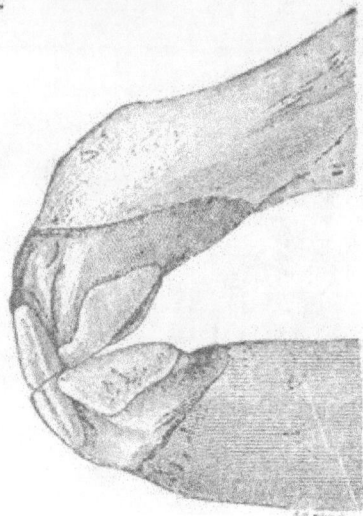

FIG. 43.

Three Months (Figs. 41, 42, 43).—From in front, the pincher teeth are seen almost entirely free, though the gums still encircle the base of their free portion. The intermediate teeth are in contact by the internal portion of their anterior border. In profile, the jaws are seen wider, thicker, and less curved; the intermediate teeth are not yet completely free. Their free borders are separated behind; the dental tables are slightly used; both borders of the intermediate teeth are slightly worn toward their inner edge. The incisive arch increases in extent transversely.

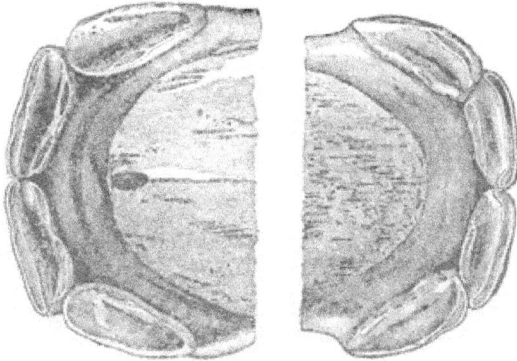

FIG. 41.

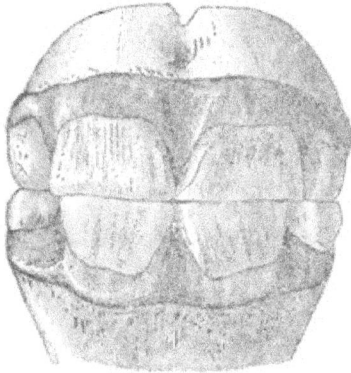

FIG. 45.

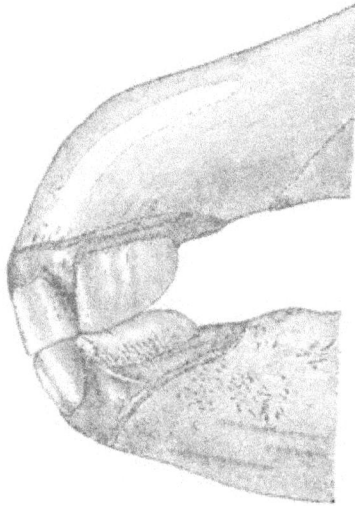

FIG. 46.

Four Months (Figs. 44, 45, 46).—From in front the transverse diameter of the jaws is increased; the intermediate teeth are further out.

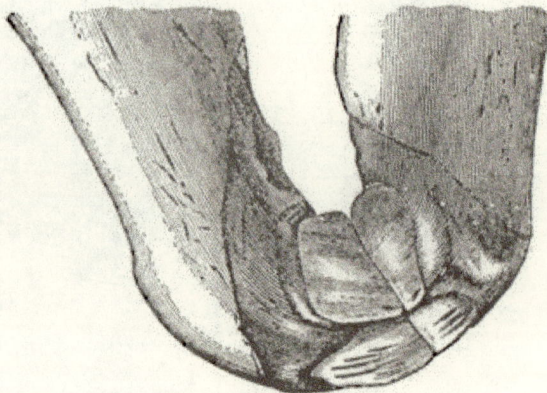

FIG. 17.

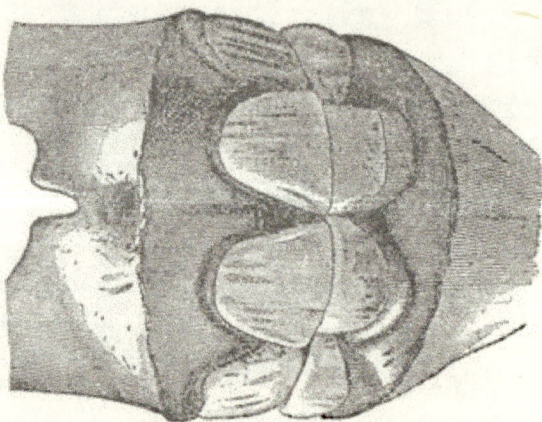

FIG. 18.

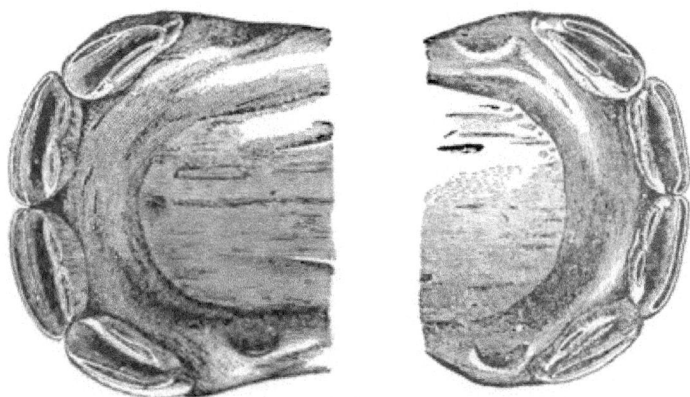

FIG. 49.

Five Months (Figs. 47, 48, 49).—The jaws are thicker and the incisive arch is wider transversely. From in front the intermediate teeth are seen almost entirely free. In profile these teeth are in contact by the whole extent of their anterior border; the table of the pinchers are more worn, especially above. In each jaw the anterior border of the intermediate teeth is worn almost for its whole length; the posterior border is less worn. Below is seen, under the mucous membrane, the internal edge of the corner teeth, which are preparing to pierce through. They form a little elevation of the mucous membranes placed immediately behind the internal border of the corresponding intermediate teeth.

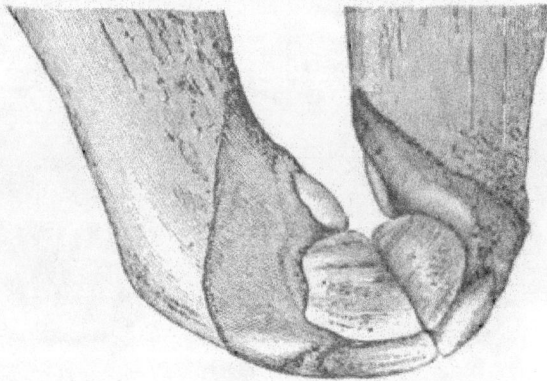

Fig. 50.

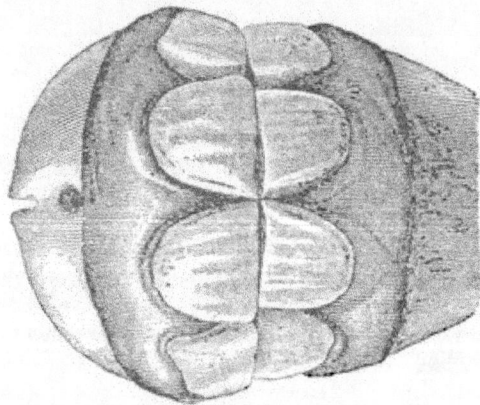

Fig. 51

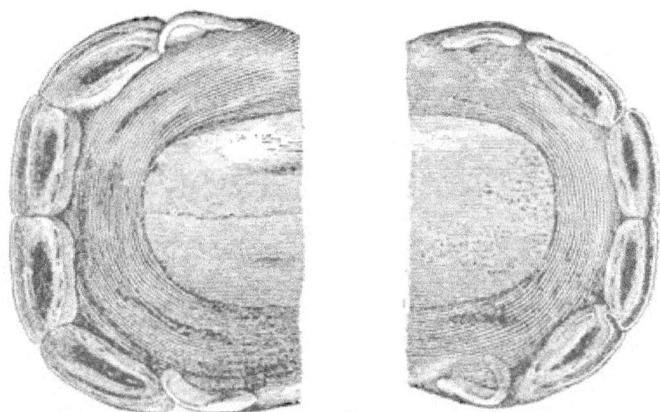

FIG. 52.

Ten Months (Figs. 50, 51, 52).—These jaws come from a pony which was rather tardy in marking its age; nevertheless, they are useful plates from which to study the diverse characters of this period. The pinchers and intermediate teeth are entirely free from the gum; the anterior face of these teeth is polished, the little striations or canals are less visible; the intermediate teeth, which have been in contact for some months, are worn on both borders. The corner teeth are just appearing, but are not yet in contact with each other. The wearing of the pincher teeth is greater than that of the intermediate teeth.

SECOND PERIOD.

About One Year.—The corner teeth have protruded from
the gums, but the inferior ones are not yet in contact with the
superior teeth. The inferior pinchers, if not leveled, are at least
very much used on both their borders. The incisive arch commences to be a little depressed in the centre. The superior pinchers
and intermediate teeth just commence to wear at their posterior
borders. (Figs. 53, 54, 55.)

About Sixteen Months.—The superior corner teeth meet the
inferior and commence to wear at their anterior border; the
necks are clear of the gums. Often at this time the inferior
pinchers are leveled, but the intermediate teeth are rarely more
than slightly worn. The incisive arch is flattened in front.
(Figs. 56, 57, 58.)

About Twenty Months.—The inferior corner teeth are nearly
leveled; the superior ones are less so. The inferior pinchers
stand out from the gums, and the intermediate teeth are often
leveled. The incisive arch becomes less convex. (Figs. 59,
60, 61.)

About Two Years.—The inferior dental arch is completely
leveled at the pinchers and intermediate teeth, and the superior
arch is nearly so. The superior pinchers stand out from the
gums, and behind them is found a moderately sensitive swelling,
—due to the permanent teeth, which are pressing on the gum of
the palatine arch. The intermediate teeth are free from the
gums above and below. The incisive arch has widened from
side to side and the pinchers and intermediate teeth form almost
a straight line. (Figs. 62, 63, 64.)

(64)

Girard thought that the pinchers were leveled at ten months, the intermediate at one year, and the corner teeth at fifteen to twenty-four months, but this is too definite; the leveling of the temporary incisors is somewhat irregular, and is considerably modified by the depth of the cups, the amount of cement which they contain, and the character of the food upon which the animal is fed. Experience will teach the observer to place much value on the condition of the corner teeth, the amount of wear of the superior incisors, and upon the color, which gradually becomes darker. At the end of this period the pinchers become broken, loose, and ready to fall from their sockets; they are less solidly fixed in the jaw and may be broken off, or are pushed out naturally by the permanent teeth, which replace them.

During this period, especially during the second year, the variations in the amount of wear of the teeth in different animals may be very marked; but, by a careful comparison of the use which each pair of teeth—pinchers, intermediate, and corners—have undergone, and, with close observation of the development and size of the bones, taking into consideration the intermaxillary bones, and the width and thickness of the body and branches of the maxilla, a very close differentiation of a month or two may be made. Especial note must be made of the amount of gum which still covers the crown of the teeth, or their freedom from the gums, the discoloration on their surface, the polishing off of the small striations, and the evidence of the protrusion of the permanent teeth under the gums behind the deciduous teeth.

5

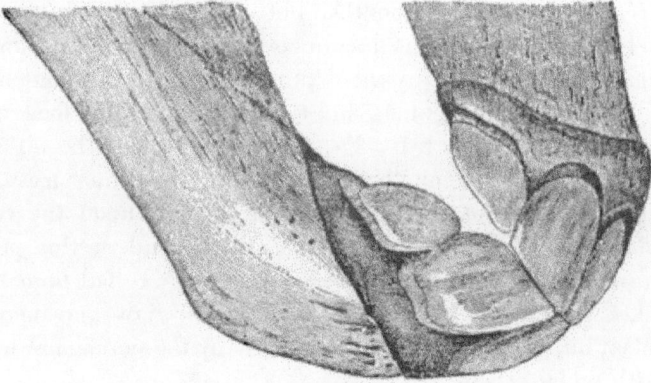

FIG. 53.

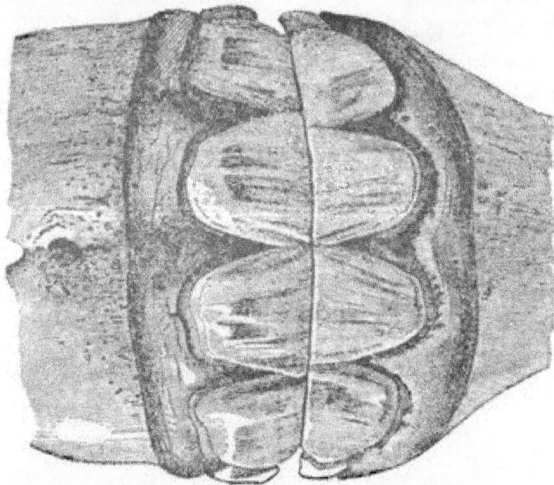

FIG. 54.

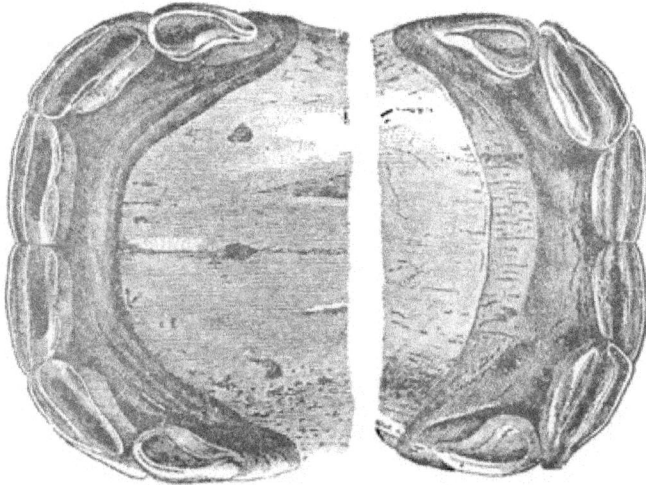

FIG. 55.

One Year (Figs. 53, 54. 55).—All of the deciduous incisors can be seen from in front; the pinchers and intermediate teeth are entirely free from the gums. In profile the superior corner teeth are not yet in contact with the inferior teeth. The tables show a decided use on the posterior borders of the intermediate teeth, which, however, is subject to variation, according to the height of the posterior border in different colts. There is generally seen at this time in the anterior border a yellow line, elongated transversely, which represents the elementary dental star. The corner teeth are still virgin. Comparison must be made between the wear of the pincher teeth and that of the borders of the intermediate teeth, and according to the amount of use which the latter have had the animal can be judged as *rising* or *off* the age indicated by the other marks.

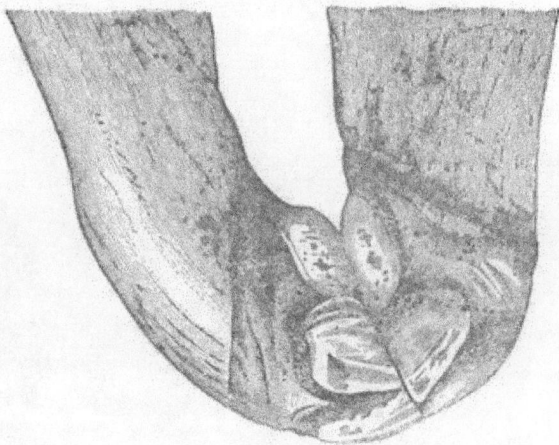

FIG. 56.

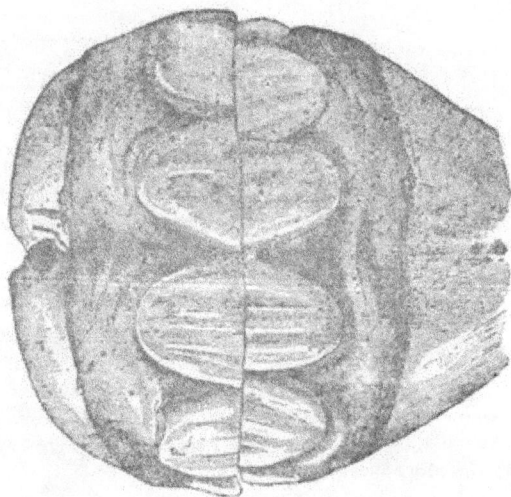

FIG. 57.

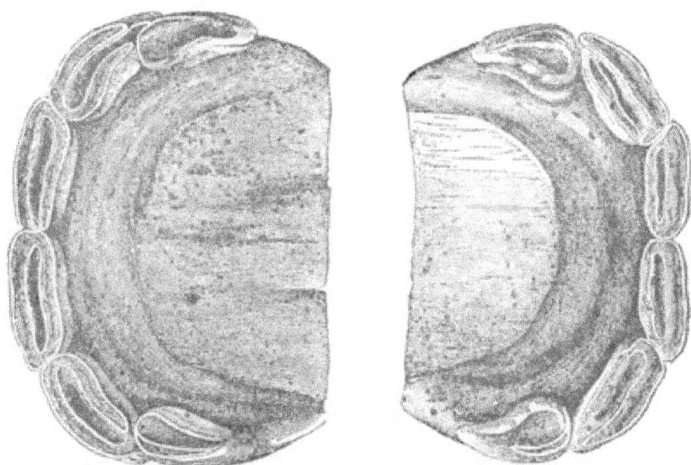

FIG. 58.

Sixteen Months (Figs. 56. 57, 58).—All of the teeth are in contact.
The superior corner teeth are in apposition with the inferior, and have
commenced to be leveled in both jaws; the enamel of the cups and the
peripheral enamel are separated; the crown of the tooth is entirely free
from the gum. Often at this period the inferior pinchers are leveled,
and sometimes the inferior intermediate teeth are also leveled. In the
upper jaw the tables of all teeth are entirely formed, the incisive arches
lose their convexity.

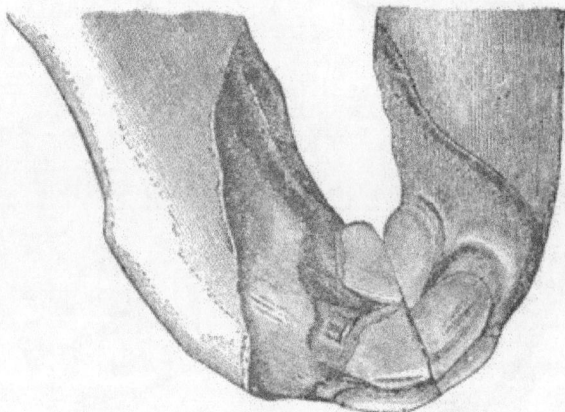

FIG. 59.

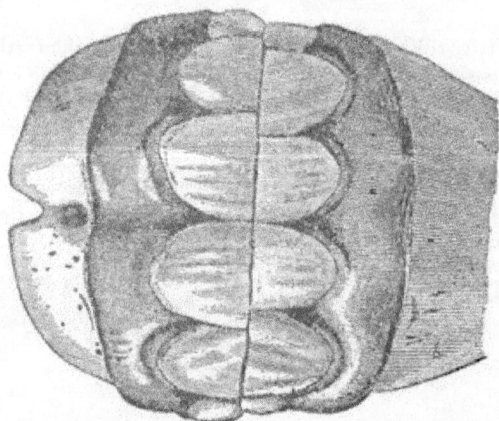

FIG. 60.

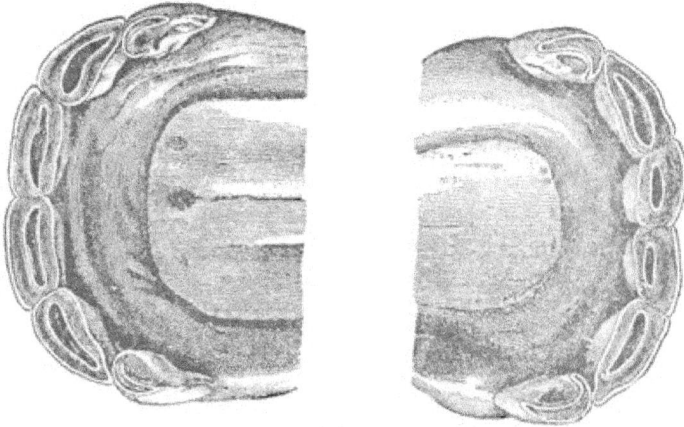

FIG. 61.

Twenty Months (Figs. 59, 60, 61).—This month comes from a thoroughbred colt, which was grain-fed from birth. It represents more use than should be at this age. The inferior corner teeth are leveled on their anterior border; the superior corners are somewhat worn, but not even comparatively to the degree of use of the pinchers and intermediates. The inferior pinchers are completely worn, and the inferior intermediate teeth are leveled; the incisive arch has become less convex, but not wide enough for a two-year-old; a bit of the gum still remains around the roots.

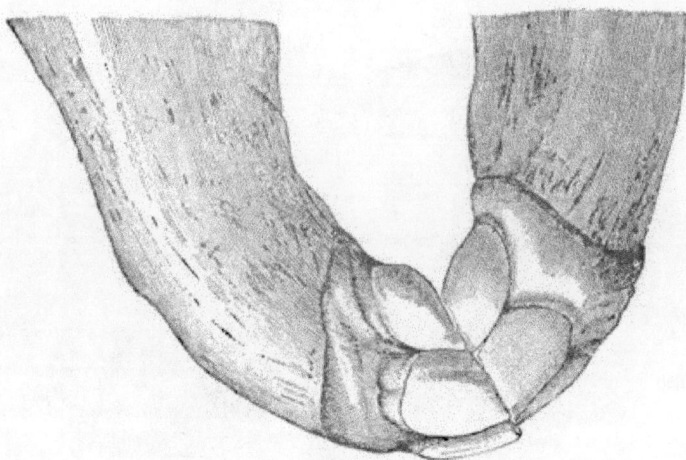

FIG. 62.

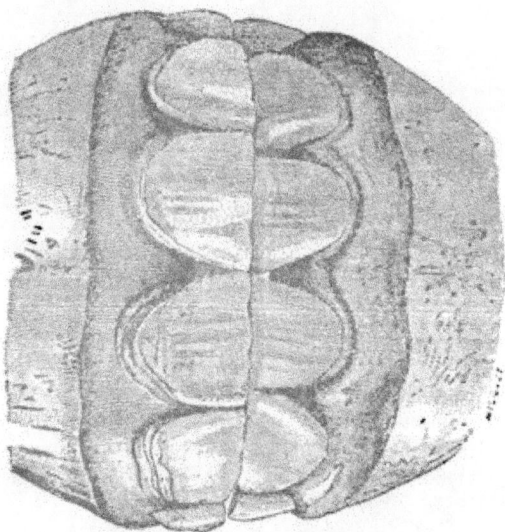

FIG. 63.

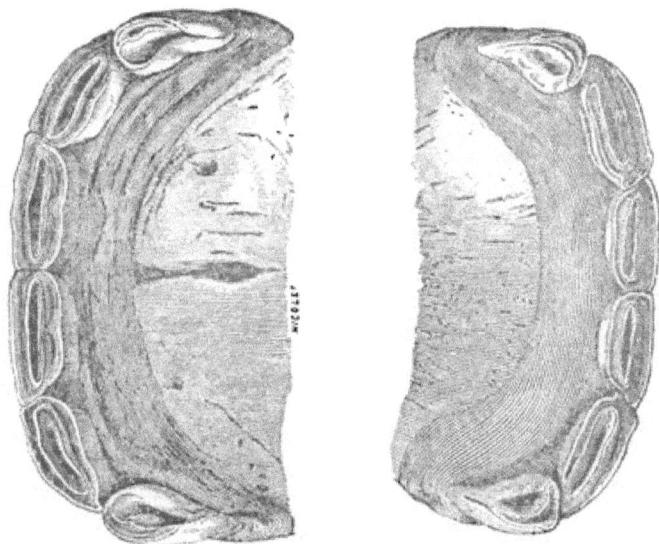

FIG. 64.

Two Years (Figs. 62, 63, 64).—This mouth comes from a common-bred colt which had no grain feed. It was two years and twenty-six days old. From in front the pinchers and intermediate teeth are seen free from the gums, indicating that they are pressed on by the permanent teeth. In profile, the corner teeth are free to their necks. The tables are well worn, and the dental stars show. The cups of the superior intermediate teeth are free from the peripheral enamel. The incisive arch is widened transversely.

THIRD PERIOD.

ERUPTION OF THE PERMANENT OR ADULT TEETH.

THIS period commences at the age of two or two and a half years, and finishes at five years.

About Two and a Half Years.—Successive falling out of the temporary pincher teeth; swelling of the gums and appearance of the anterior borders of the permanent pinchers. Ordinarily these teeth appear in the superior jaw first, and the piercing of the gums is completely effected in about six weeks to two months.

Rising Three Years.—As the colt is approaching the age of three years, we find in the superior jaw the permanent pinchers wholly out of the gums and almost reaching the level of the temporary intermediate teeth. In the inferior jaw, the borders and sometimes a greater extent of the free extremities of the teeth have appeared through the gums, though the teeth are still virgin; the intermediate temporary teeth are free from the gums at their neck, and very much worn. The corner teeth are worn so that they touch each other by their external borders. (Figs. 65, 66, 67.)

Three Years.—At three years of age all of the permanent pinchers are out of the gums and have reached the level of the temporary teeth. The permanent pinchers are wider transversely, square, and darker in color than the temporary teeth, and show little gutters on their anterior face. They differ distinctly from the others, which are smaller, more convex, have a constriction in the neighborhood of the gums, are whiter, and are not marked with gutters.

(74)

The time of the year and the race of the animal must be taken into consideration in determining the completion of three years. The better breeds of horses attain that age in mid-winter, while those of more common races do not attain it until the months of March, April, or May.

Three Years Off.—When the colt is several months from three years of age the permanent teeth are well used on their borders and in contact with each other, but the dental cups are not yet complete circles, as the enamel which forms them is still connected with the peripheral enamel toward the borders of the teeth. The intermediate teeth are very much worn, protrude from the gums, and are sometimes broken and ready to fall out. The tables of the corner teeth have become very much larger and almost cover the external borders of the teeth. (Figs. 68, 69, 70.)

Rising Four Years.—Eruption of the permanent inter-mediate teeth and progressive falling out of the temporary inter-mediate teeth mark this period. The permanent intermediate teeth appear, but have not yet reached the level of the tables of the corner teeth, and are not yet worn. The central enamel in the pincher teeth surrounds the dental cup, which is flattened from in front to behind, and is almost distinct. The corner teeth commence to be free at their necks from the gums. (Figs. 71, 72, 73.)

Four Years.—Each jaw shows four permanent teeth, with their tables on the same level; the intermediate teeth are worn both on their anterior and posterior borders, but the dental cups are not entirely separated from the outside peripheral enamel. Often the inferior pincher teeth are leveled, especially in well-bred horses. The temporary corner teeth stand out from the gums and are completely worn. (Figs. 74, 75, 76.)

Four Years Off.—Loosening and successive falling out of the temporary corner teeth, which are worn to stumps, scarcely fastened in their alveolar cavities. Sometimes one or more of the corner teeth have fallen out, and we find the inferior borders of the permanent corner teeth appearing first, more frequently in the upper jaw. The pinchers and intermediate teeth are well worn. At this period we frequently find anomalies in the eruption of the teeth. It is not rare to see the intermediate and corner teeth appear at the same time; so that an animal which is only four or four and a half years of age may have the teeth which ordinarily indicate five years. (Figs. 77, 78, 79.)

Rising Five Years.—The four temporary corner teeth have fallen out and are replaced by the adult teeth. These last have not yet reached the level of the intermediate teeth and are not yet worn. The pinchers are leveled; their central enamel, elongated from side to side, is found farther and farther away from the anterior border of the dental table. The tables of the intermediate teeth are distinctly formed. (Figs. 80, 81, 82.)

Five Years.—The mouth is complete; the incisive arch is semicircular and regular in shape; all of the permanent teeth have reached the same level. The anterior borders of the corner teeth are completely worn. The posterior are not yet worn. (Figs. 83, 84, 85.)

Five Years Off.—The above characteristics are more distinct. The age has been more marked by the continual friction and amount of work to which the corner teeth have been subjected. In the superior incisive arch, the posterior borders of the corner teeth rarely commence to be worn. The profile of the incisives shows a regular one-half circle, convex from above to below.

During this period, besides the causes already referred to which produce hasty or tardy eruption of the permanent teeth, other influences may act. Traeger, veterinarian at Doehlen, noticed that gestation in the young mare delayed the eruption of the permanent teeth; he also observed that continuous pregnancies diminished the wearing of the teeth. During this period the tusks may make their appearance, or they may be delayed until after all of the incisors are in place. In cases of doubt, a further examination should be made as to the con-

dition of the molars. By reference to the table of the eruption
of these teeth it will be seen that the first permanent molar
appears between the thirtieth and thirty-second months, the
second at about three years, and the third at four to five years.
The same rule as given for the second period applies here also,
and a close comparison should be made of the comparative
wearing of the pincher, intermediate, and corner teeth, as the
first or second pair may be advanced in use and not correspond
to the freshness of the delayed later teeth.

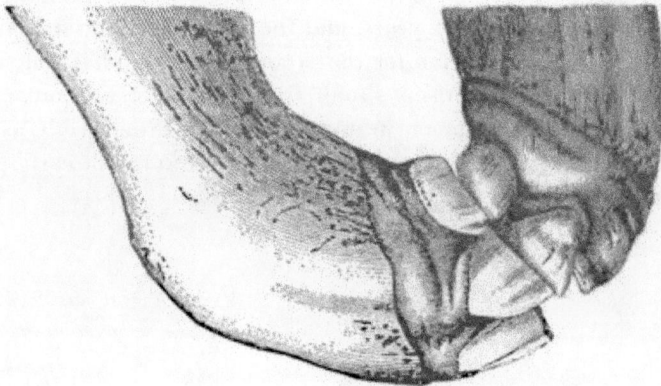

FIG. 65.

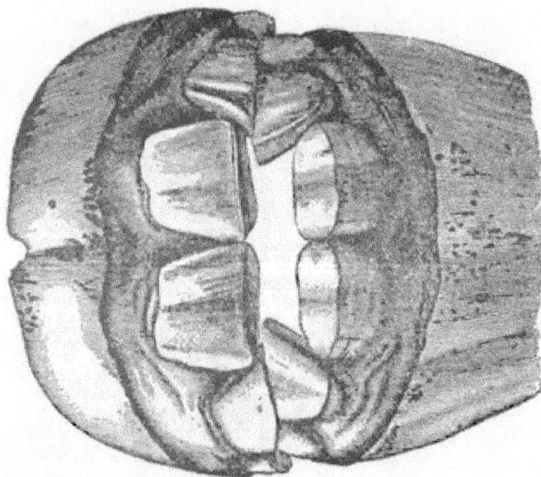

FIG. 66.

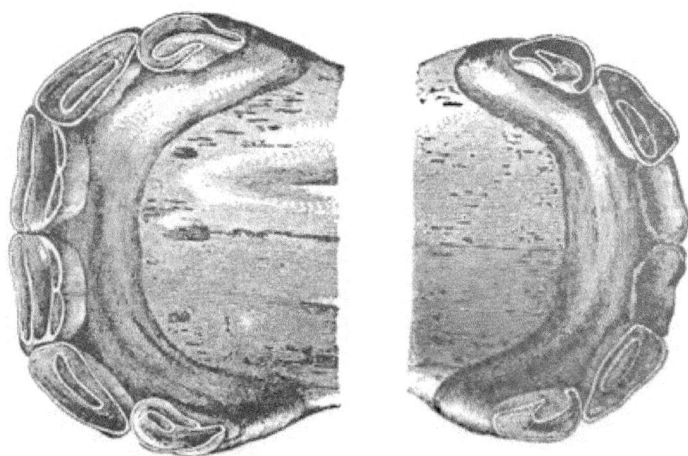

FIG. 67.

Rising Three Years (Figs. 65, 66, 67).—From in front, in the upper jaw, the two permanent pincher teeth are seen not yet opposite to the level of the intermediate deciduous teeth; below, the adult pinchers are just coming through the gum, showing a small portion of their anterior face. In profile, the intermediate teeth are very much worn, and the constriction on their neck is pushed out beyond their gums; the corner teeth are much shortened; the dental table shows slight wear of the superior pinchers, which has been produced by the eruption of these teeth before the inferior temporary pinchers had fallen out, and, consequently, they have worn against the latter. The intermediate teeth are completely leveled; the corner teeth are much used.

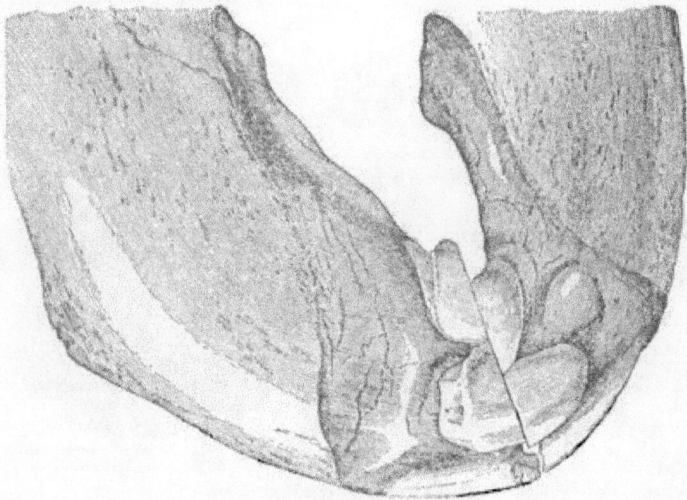

FIG. 68.

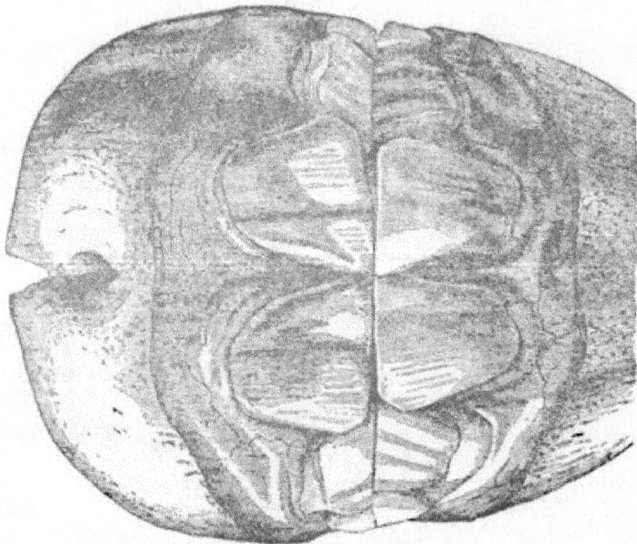

FIG. 69.

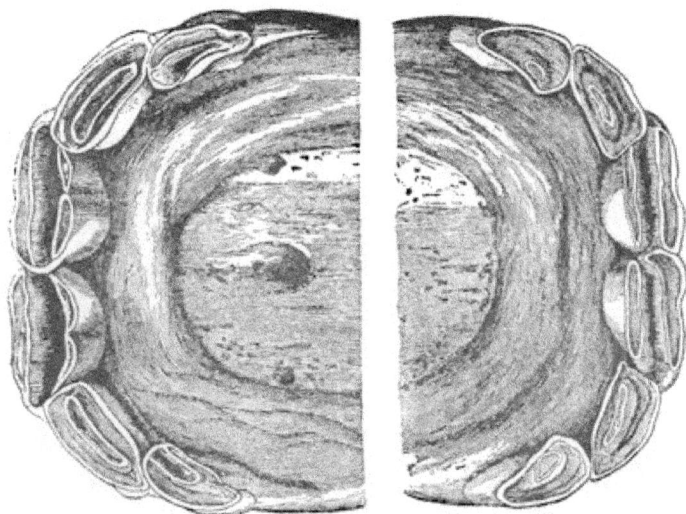

FIG. 70.

Three Years Off (Figs. 68, 69, 70).—From in front the four permanent pincher teeth are seen, much larger and stronger than the
neighboring teeth. The anterior borders of the superior pinchers are
oblique, and their external borders are not yet in contact with the corresponding part of the inferior teeth. In profile, the intermediate teeth
are seen much used; the corner teeth are short, and show the constriction at their neck. The tables are worn off level. Between the corner
and the intermediate teeth on the left is seen the protrusion of gum
made by the permanent intermediate tooth which is shortly to appear.
The dental tables of the inferior intermediate teeth are very much worn,
the superior teeth somewhat less so. The inferior corner teeth are
entirely leveled. In this mouth the tables of the inferior pincher teeth
are the most worn, as these teeth came out before the superior ones did.

6

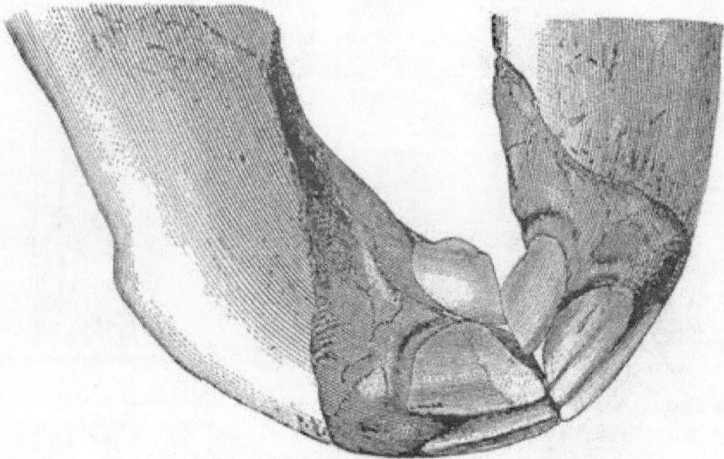

FIG. 71.

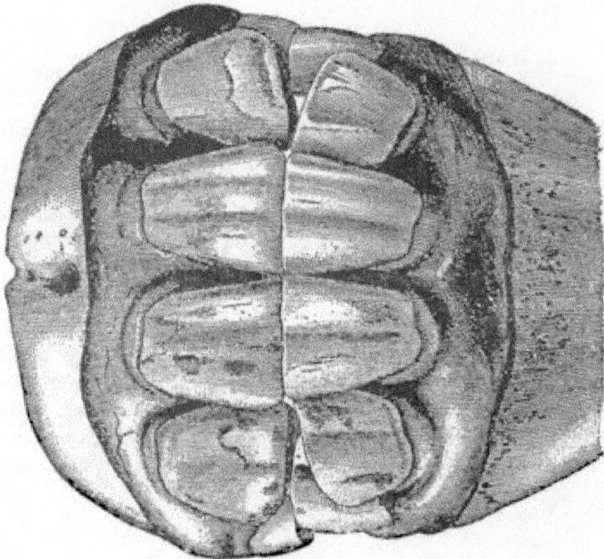

FIG. 72.

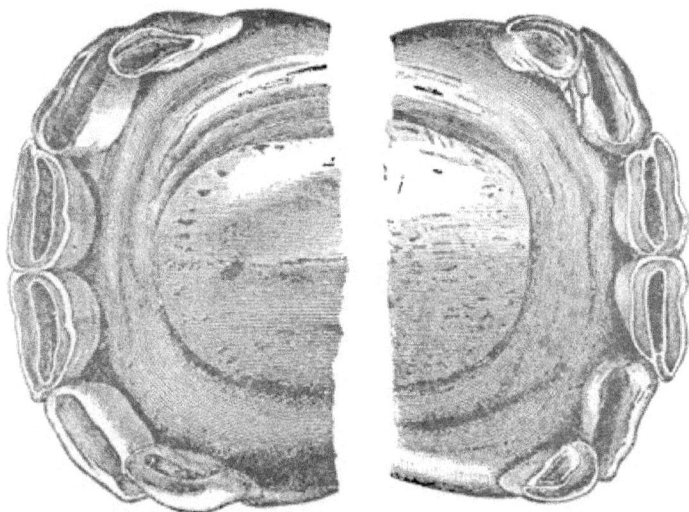

FIG. 73.

Rising Four Years (Figs. 71. 72. 73).—From in front are seen the eight adult incisors ; the pinchers in contact with the opposite ones, and intermediate teeth not yet out to the level of the pinchers. In profile, this is also seen in this month. The corner teeth are very much worn ; the tables of the pinchers are considerably worn, making almost a complete separation of the central, or cup, enamel from the peripheral enamel. This advanced wear of the pincher teeth is not in direct harmony with the amount of use of the corner teeth.

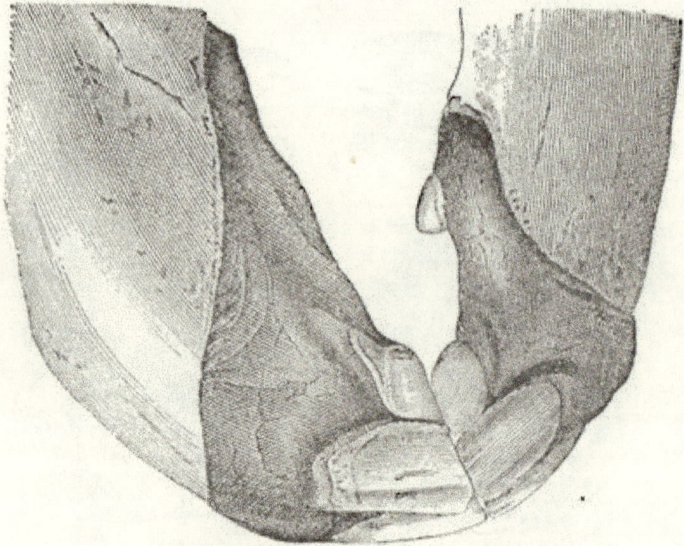

FIG. 74.

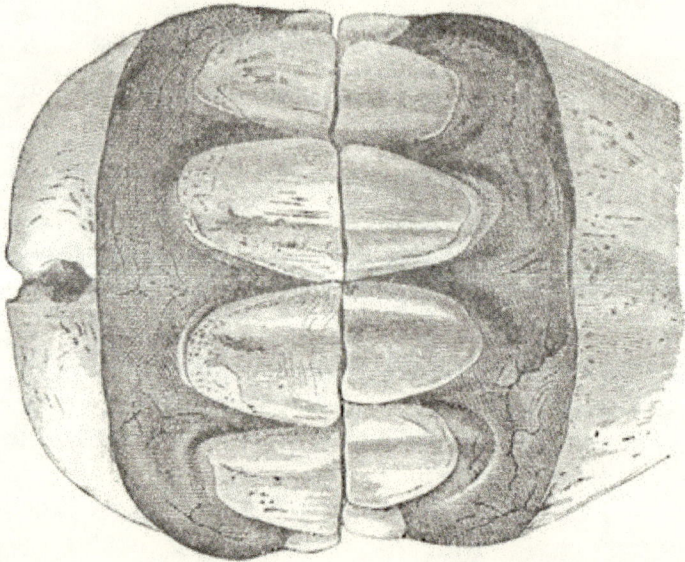

FIG. 75.

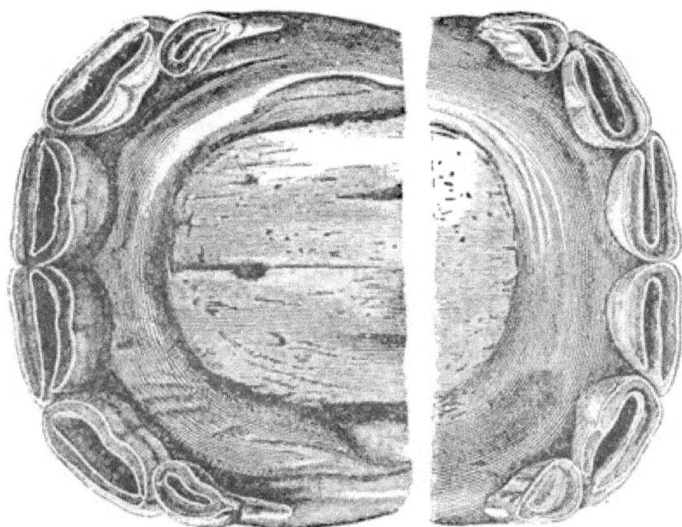

FIG. 76.

Four Years (Figs. 74, 75. 76).—From in front the four superior permanent teeth are seen in contact with the inferior teeth. The jaw has attained such a width that the corner teeth are almost hidden. In profile, the latter are seen to be very small. The superior ones have commenced to be pushed out from the jaw. In the lower jaw are seen the tush teeth. The tables of the intermediate teeth are much worn, especially in the upper jaw, in which the eruption took place first. The central enamel is only separated in the superior left-hand teeth ; the inferior corner teeth are almost leveled, the superior ones completely so. The latter are being pushed out, and show their roots.

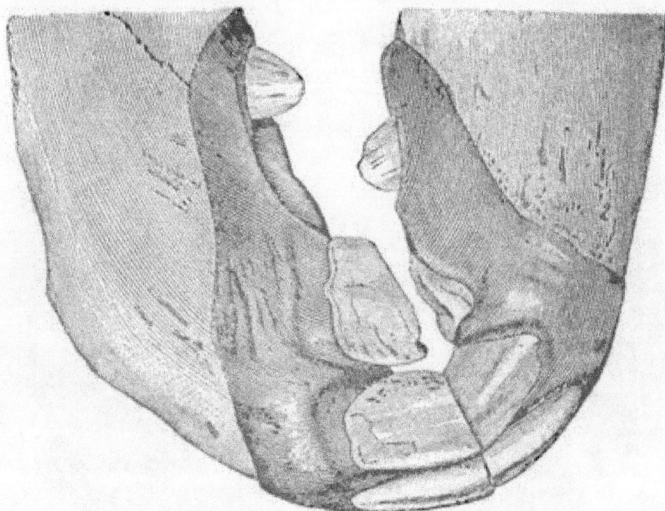

FIG. 77.

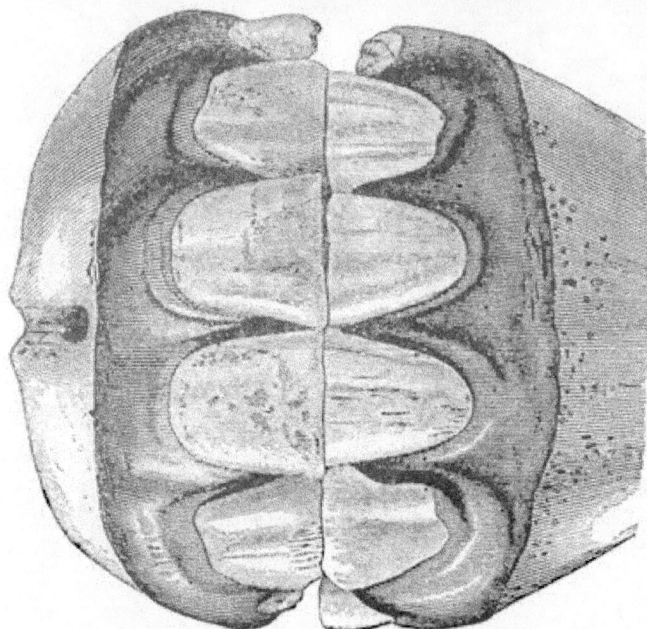

FIG. 78.

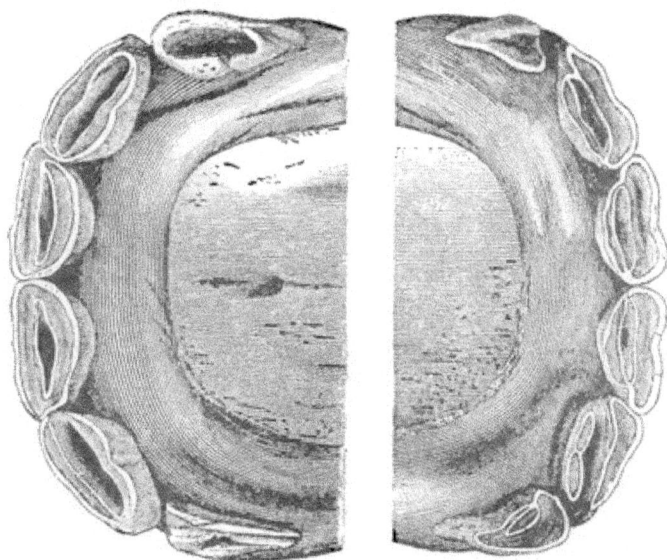

Fig. 79.

Four Years Off (Figs. 77, 78, 79).—In front the intermediate permanent teeth are seen in contact with each other; the inferior and superior left-hand permanent corner teeth have appeared. In profile, it is seen that these teeth have not been completely pushed through the gums. The tush teeth have appeared. The right-hand superior milk-tooth is ready to fall out; nothing remains but its roots. The inferior tooth on the same side is leveled, but still firmly imbedded in the jaw. The superior intermediate teeth, which preceded the eruption of the inferior teeth, show considerable wear. The cups have formed in the pincher teeth.

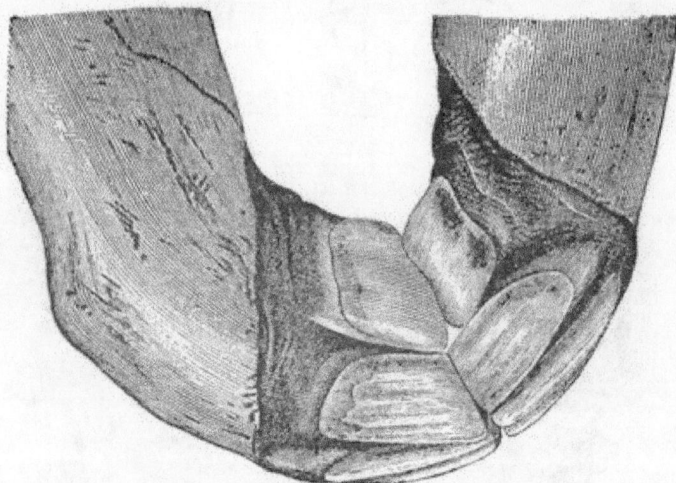

FIG. 80.

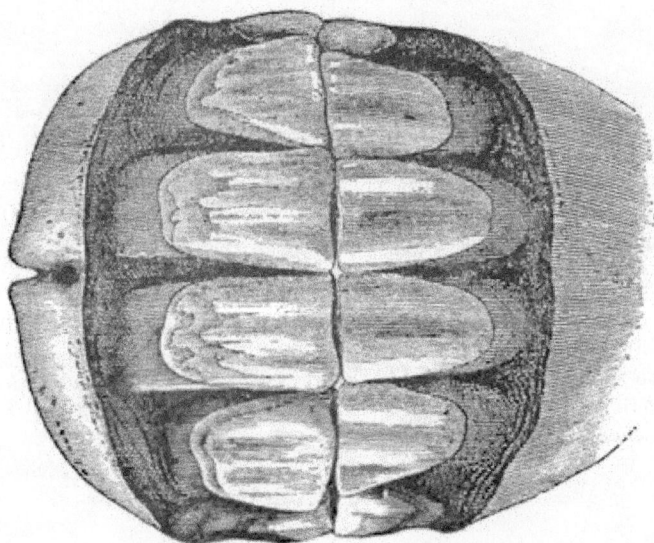

FIG. 81.

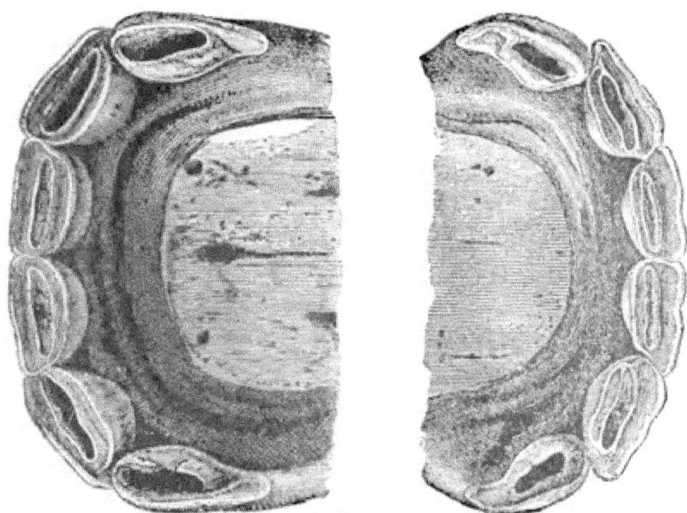

FIG. 82.

Rising Five Years (Figs. 80, 81, 82).—The four corner temporary teeth have been replaced by the permanent teeth, but these are not on a level with the intermediate and are entirely virgin. The tables of the other teeth show a more decided use than those in the last figures. Above, the cups are formed in both the pincher and intermediate teeth; the cups are nearly formed in the inferior intermediate teeth.

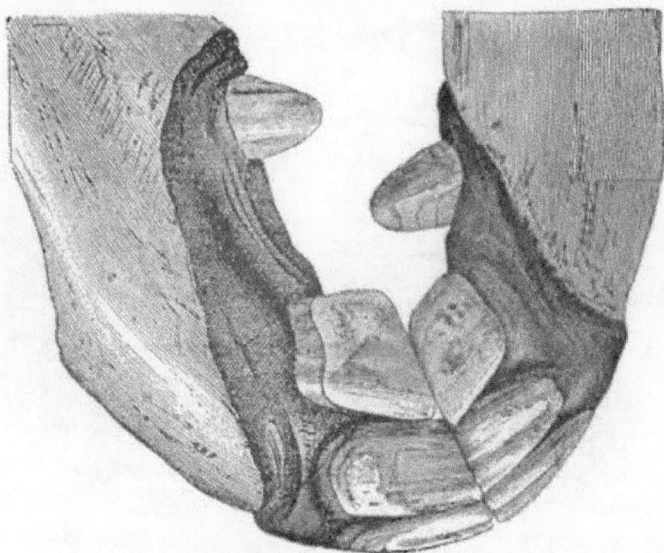

FIG. 83.

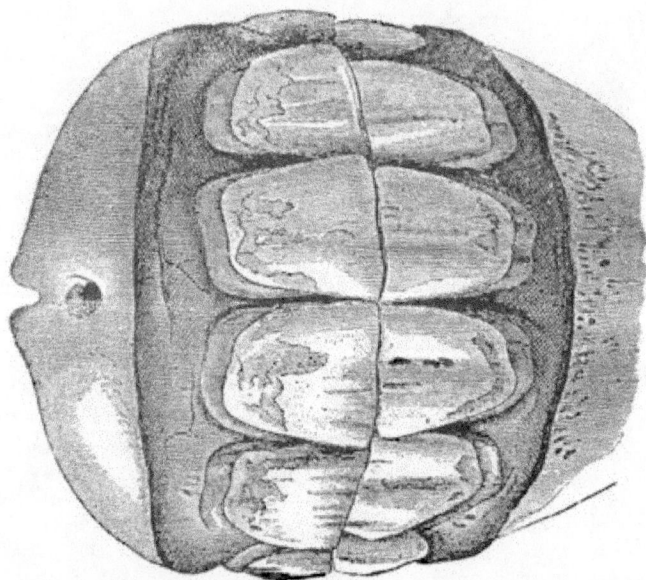

FIG. 84.

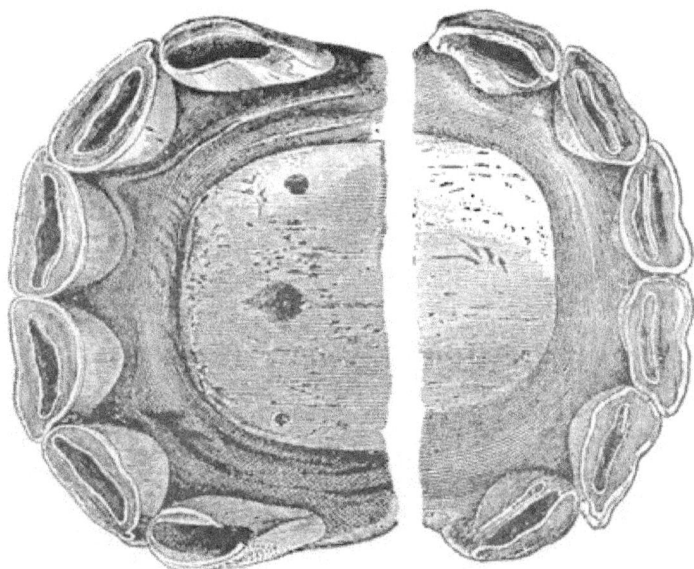

FIG. 85.

Five Years (Figs. 83, 84, 85).—The mouth is entirely made; all of the permanent teeth have reached the same level in both jaws. The jaws' are convex in both directions; the tush teeth have completed their eruption. The corner teeth have commenced to wear on their anterior border; the pincher teeth are leveled, but the cups are still elongated from side to side and are very narrow; they commence to approach to the posterior border of the teeth. This form of cup indicates that in these teeth the cups are very shallow. The incisive arches form a regular half-circle.

FOURTH PERIOD.

LEVELING OF THE PERMANENT INCISORS.

DURING this period the signs furnished by the dental apparatus become more difficult to recognize, and the determination of the exact age is less precise than at an earlier period. The points to be examined from six years are, first, the wearing of the corner teeth, the form of the transverse diameters of the teeth, the position of the central enamel on the surface of the table, and the general outline of the incisive arch.

At six years, after the complete leveling of the teeth, the tables of the pinchers commence to become oval in shape. The posterior border of the corner teeth has become worn and the cup is completely separated from the peripheral enamel. The cement disappears from the anterior face of the pinchers, so that they become whiter in color. Very frequently at this age the notch commences to show on the outer border of the superior corner teeth.

At seven years, all of the teeth are denuded of cement on their outer faces and are much whiter in color. The inferior corner tooth, which is narrower from in front to behind than the superior, wears only the anterior portion of the latter and makes distinct, in profile, the notch which we have just seen often commences at six. The pincher teeth become narrower from side to side, and the oval of the posterior border shows a projection backward; the intermediate teeth become oval, and, in both, the cups become decidedly narrower and are found nearer the posterior border of the teeth. In profile, the incidence of the jaws has altered from the convex arch of six years and tends to assume the form of an ogive.

At eight years, the incisors commence to turn from a white to a yellowish white. The arches are narrower from side to

(92)

side, the obliquity of the teeth is greater, and, as the inferior corner teeth now commence to wear by their posterior borders, their tables become elongated and correspond to the tables of the superior corner teeth in size, so that the notches of the latter become less marked. The interspaces between the teeth become more marked and the gum commences to retract from the crowns, giving a square cut-off appearance. The tables become more convex on their posterior borders, and the cups, which continually approach the posterior borders of the teeth, in the pincher teeth become convex behind. In front of them, half-way to the anterior border of the teeth, a transversely elongated dark-yellow line appears. This is the *dental star*, brought into view by the uncovering of the calcified dental pulp. If the structure and formation of the incisive teeth are remembered, the exact value of the dental star will be better appreciated. In the virgin tooth the upper extremity is hollowed by the cup, and the dental pulp occupies the space between the anterior face of the cup and the anterior face of the tooth itself. But we have seen that the cup, pointing downward, also inclines toward the posterior border of the tooth. At eight years the cup is at the division of the posterior and middle thirds of the table of the tooth, and the dental pulp is found on the line between the middle and anterior thirds. The transverse extent of the dental star is much less in the intermediate and corner teeth than it is in the pincher teeth.

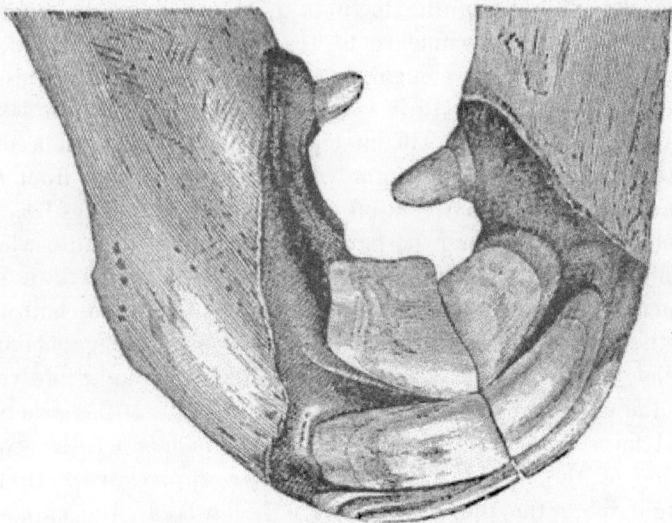

FIG. 86.

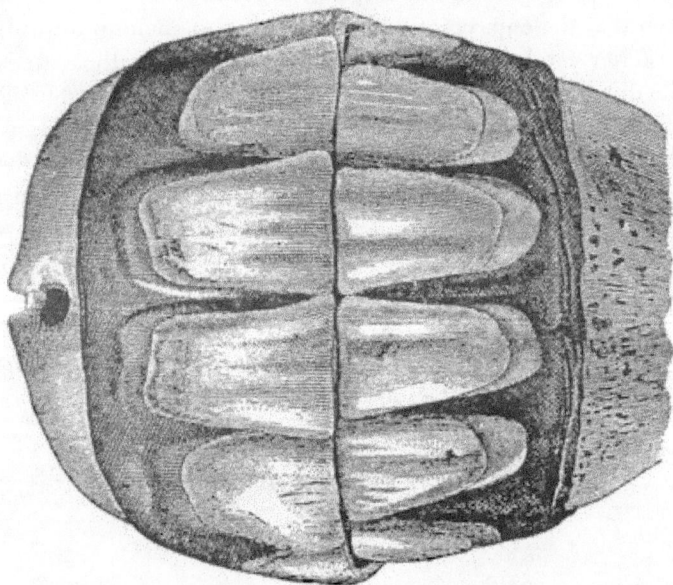

FIG. 87.

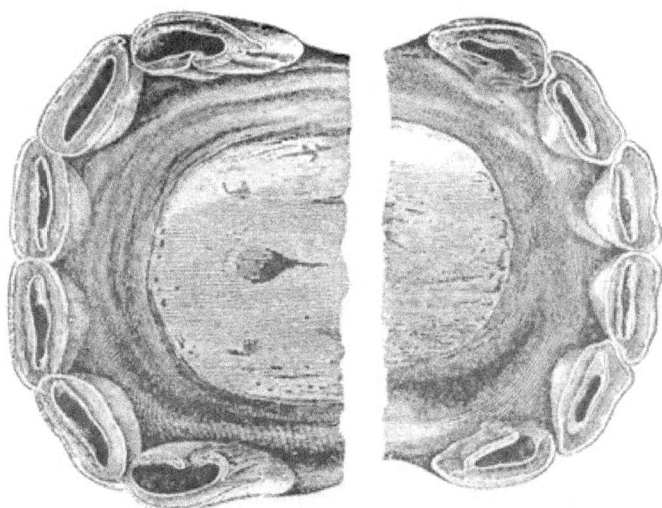

Fig. 88.

Six Years (Figs. 86, 87, 88).—In front, the teeth appear much as they were at five years of age. In profile, we see in this animal a more tardy eruption of the tush teeth, which are not yet quite free from the gums, and are, therefore, of little value as regards the age. The tables furnish a most accurate guide; the posterior border of both inferior and superior corner teeth are worn; the pincher teeth are leveled and their tables tend to an oval form. It is seen, however, that the inferior cups are thicker at their anterior borders,—due to a small portion of the surface enamel still remaining. The cups are narrower from side to side than at five years, and somewhat closer to the posterior border of the table; the same appears on the intermediate teeth. It will be noticed that the cups of the superior corner teeth have fissures on their posterior borders, which is of frequent occurrence and does not interfere with judging the amount of work which they have performed.

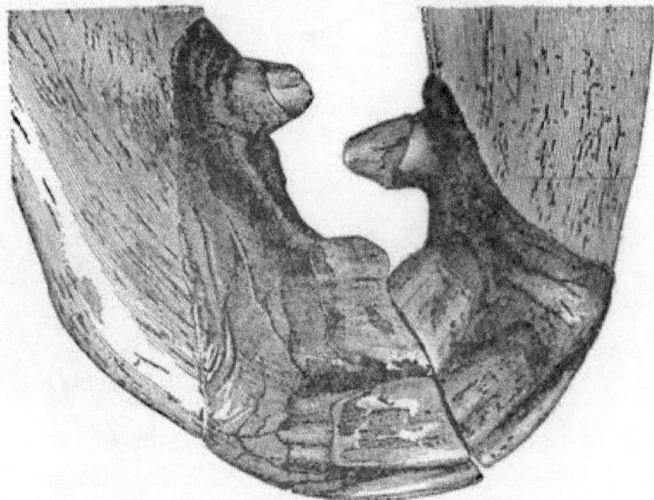

FIG. 89.

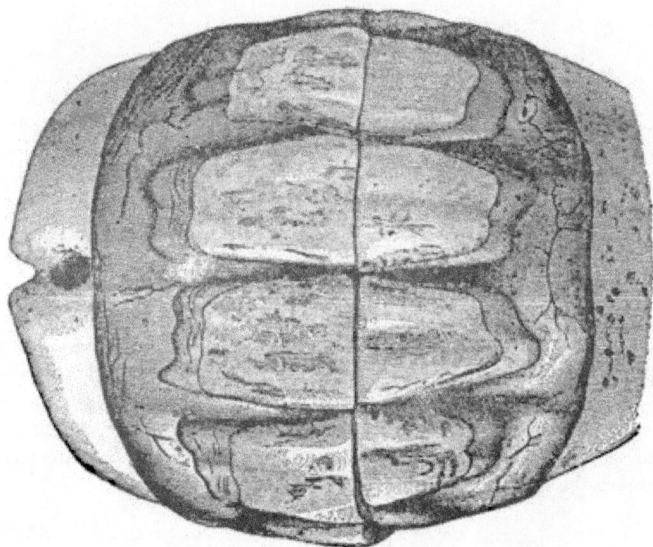

FIG. 90.

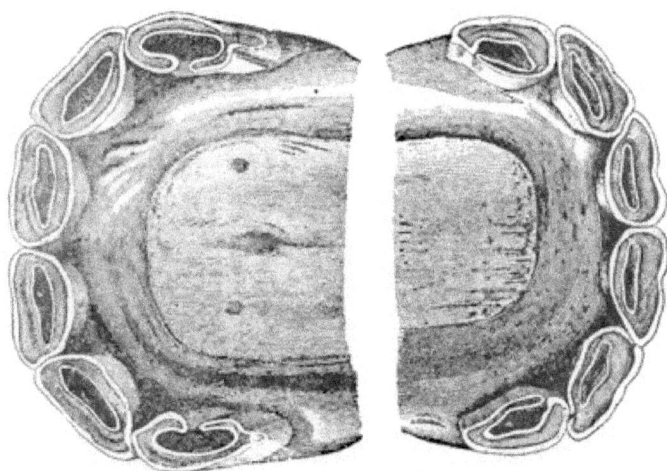

FIG. 91.

Seven Years (Figs. 89, 90, 91).—There is nothing special to be seen in front except that the teeth are whiter,—due to the disappearance of the cement, which has been worn from the surface of the enamel. In profile it is seen that the table of the inferior corner tooth is narrower than that of the superior from the front to behind, so that a notch is formed on the posterior corner of the latter. The incident of the tooth is less perpendicular than at six years. The cups of the tables of the intermediate teeth are wider from in front to behind and narrower from side to side. In the corner teeth the wearing surface is larger and the cups are smaller. The pincher teeth are oval and the intermediate teeth commence to become so. In this month, again, the superior corner teeth are fissured on their posterior borders.

7

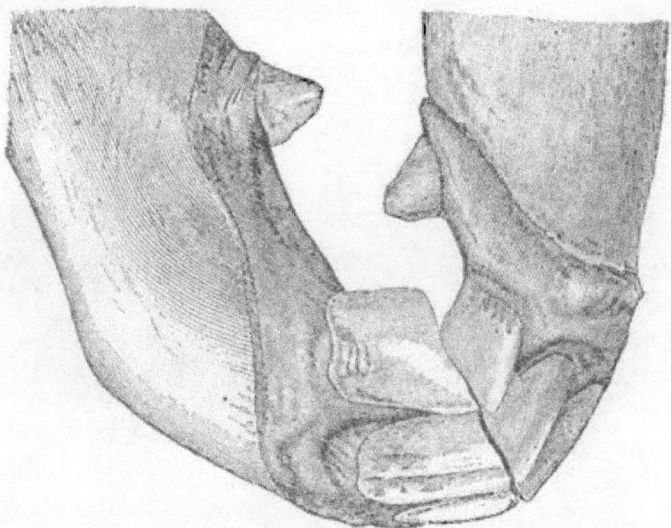

FIG. 92.

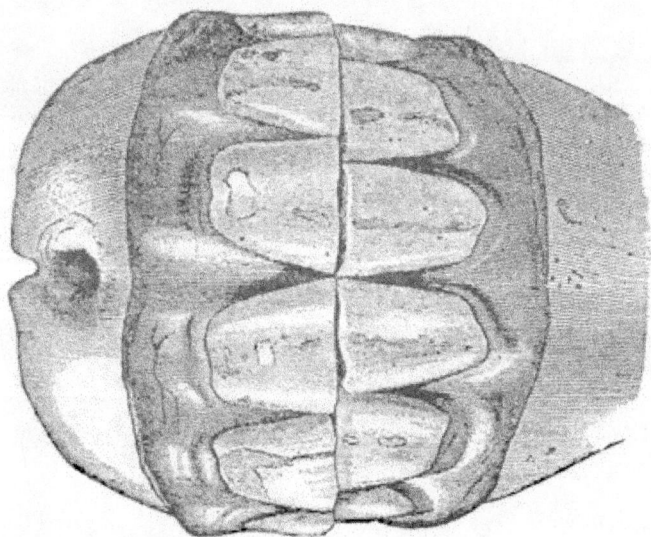

FIG. 93.

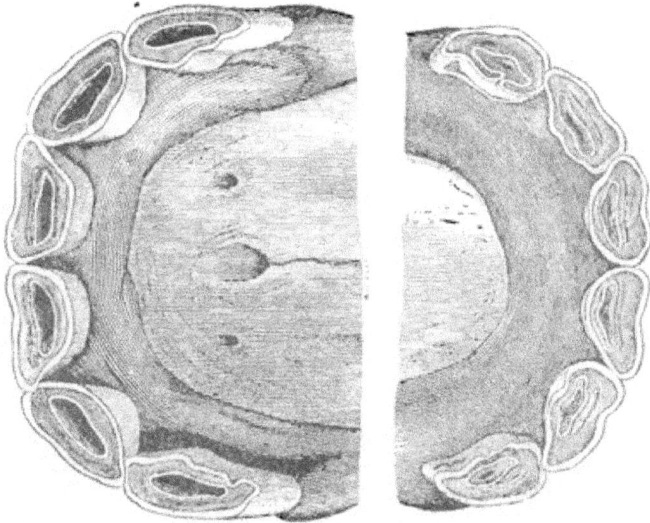

FIG. 94.

Eight Years (Figs. 92, 93, 94).—The direction of the incisors is decidedly changed; the inferior and superior arches are opposed obliquely; seen from in front, the teeth project at the line of apposition. In the profile this is more apparent, and the arch assumes more the form of an ogive. The incisive arches are still regular, but decidedly smaller than the earlier ages. All of the inferior tables are leveled; the pincher and intermediate teeth are oval, the corner teeth begin to become so. The cups commence to assume an angular form behind, and are narrow. The dental star has appeared in the pinchers and commences to show in the intermediate teeth, between the anterior border of the table and the corresponding part of the cup.

FIFTH PERIOD.

AFTER this time, which is commonly known as "past mark of mouth," commencing at nine years and extending to old age, the regularity of the wearing of the teeth and the certainty of the signs of age furnished by them become more variable. The uncertainty increases greatly after fifteen years, and a year or two later the estimation of age from the teeth can only be a conjecture, based upon experience and subject to error of one, two, or even several years, which becomes greater the older the animal is.

The changes of this period are: The successive alteration in shape of the tables of the teeth; the position of the cups in the incisors of both jaws; the form and location of the dental stars in the tables; the obliquity or degree of incidence in the incisive arches; the convergence of the crowns and the narrowness of the jaw holding the roots; the thickness of the enamel on the anterior and posterior faces of the teeth; the appearance of the cement around the roots; and, the shape of the bones of the face and jaw.

At nine years, the pinchers are round; their cups are triangular and their dental stars are more distinct, but narrower. The intermediate teeth commence to become round and the corners oval; the superior pinchers are often leveled; the notches on the superior corners often disappear.

At ten years, the tables of the pinchers are decidedly round; the cups are very small and distinctly triangular. The intermediate teeth assume the shape of the pinchers the previous year. The dental star is nearly in the middle of the table.

At eleven years, the corner teeth are rounded. The cups are only small spots near the posterior borders of the tables; the dental stars are in the middle of the tables. The inferior

(100)

corners are as large at the gums as at their free extremities, and notches re-appear on the superior corner teeth.

At twelve years, all of the teeth are round; only a trace of the cups remains in the inferior ones. The superior corners are leveled. Both incisive arches are much narrower and the tongue shows over their borders. The inferior border of the jaw-bone becomes sharp, and a flattening of the sides of the face, over the roots of the superior molars, is seen. The incidence of the incisive arches increases, especially if the teeth are unusually long.

At thirteen years, the signs of twelve years are more marked. The notch on the superior corner teeth is greater. The cups usually disappear from the inferior incisors at this age, and the superior pinchers become round.

At fourteen years, the pinchers become triangular; the incisive arch is depressed in front and becomes decidedly narrower.

At fifteen years, the intermediate teeth commence to become triangular. The dental star is round in all the lower teeth and is dark and distinct. The cups of the upper teeth are smaller in size.

At sixteen years, the intermediate teeth are triangular.

At seventeen years, all of the lower teeth are triangular, and the dental stars are small and round. The cups of the upper corner teeth have disappeared and those of the others are round.

After nineteen, the cups have usually all disappeared, the teeth approach a line parallel to the bones of the jaw, which is especially marked in the lower jaw, and the arches are flattened from side to side. The lower teeth may be worn almost to the gums, and deposits of cement around their roots may develop to supply a wearing surface.

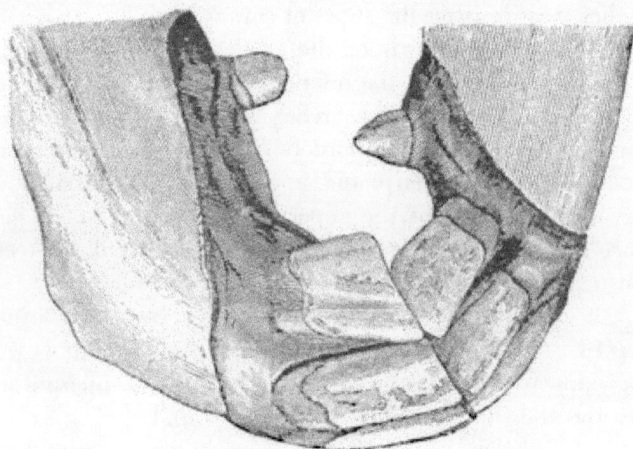

FIG. 95.

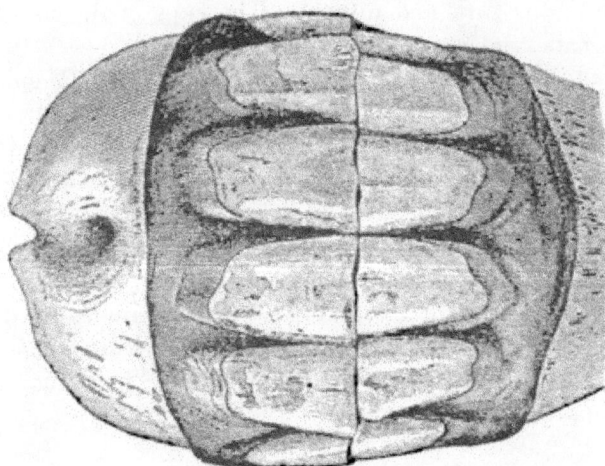

FIG. 96.

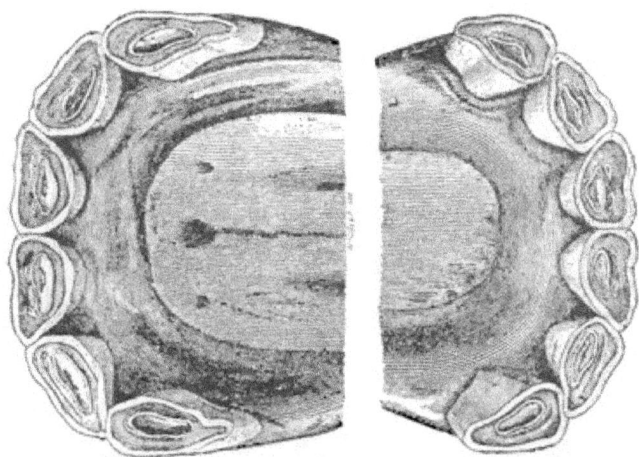

FIG. 97.

Nine Years (Figs. 95, 96, 97).—There is no special change to be seen from in front or in profile. although ordinarily the teeth are more obliquely formed and less fresh-looking than at eight years of age. The notch on the superior corner has generally disappeared ; the tables, however, at this age are characteristic ; the pinchers are rounded and their cups have assumed a triangular form ; the dental star is narrower and more distinct, and is nearly in the centre of the table ; the intermediate teeth commence to become round, and the corner teeth are oval ; the superior pincher teeth are sometimes leveled ; the inferior incisive arch is narrower and depressed in the centre.

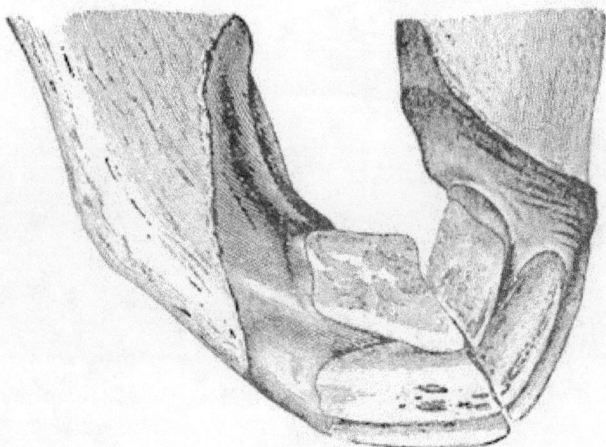

FIG. 98.

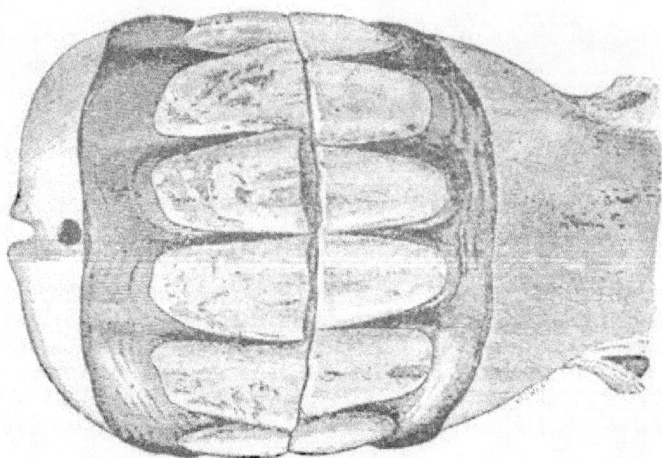

FIG. 99.

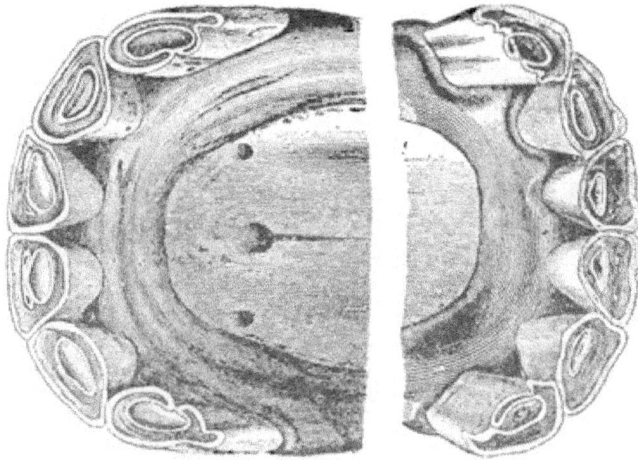

FIG. 100.

Ten Years (Figs. 98, 99, 100).—From the increased obliquity of the teeth the jaws become prominent in front, and at the ordinary height of the head the inferior incisors are hidden when looked at from in front. The ogive formed by the two sets of incisors, when looked at in profile, is more closed; the angle of the teeth increases, and a greater interspace is found between the intermediate and corner teeth. On the tables the inferior pinchers are seen to be more rounded than at nine years, and the cup is smaller and decidedly triangular; it is closer to the posterior border of the teeth. The dental star, still more apparent, is found about the centre of the teeth. The intermediate teeth are rounded and the corner teeth commence to assume this form. The example furnished by the plate shows deep fissures on the posterior border of the corner teeth. The incisive arch is flatter in the centre. In the plate showing the teeth from in front the superior pinchers are seen to be somewhat worn from cribbing.

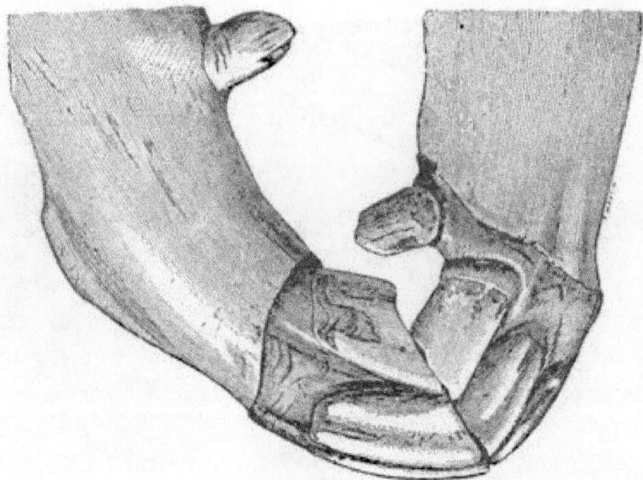

FIG. 101.

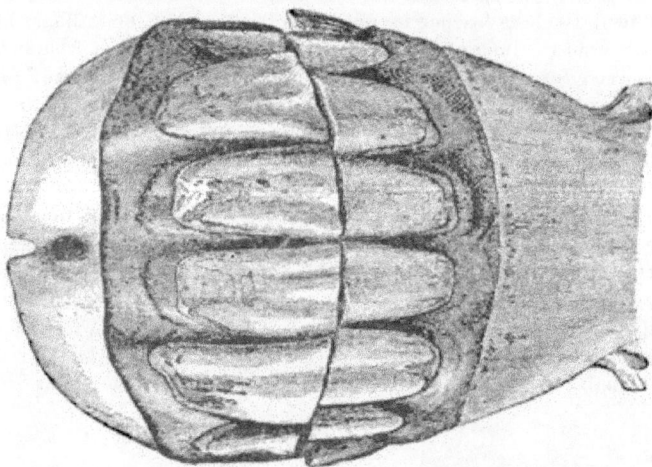

FIG. 102.

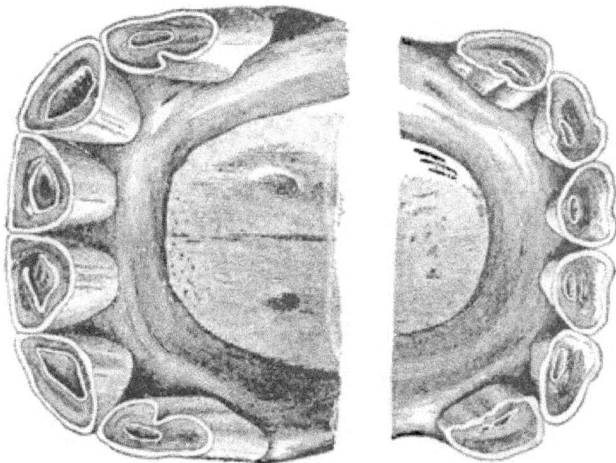

FIG. 103.

Eleven Years (Figs. 101, 102, 103).—The incidence of the jaws increases in obliquity so that the head of the horse must be raised to see both from in front. In profile, the superior corner tooth is decidedly more oblique than the intermediate one ; the inferior is the same size at its free surface and its base, and its gum is square. The inferior tables are round on the intermediate and rounded on the corners. In the lower jaw the cups are diminished to little islands close to the posterior borders of the tables, and the dental stars are narrower from side to side and near the centre of the tables. In the upper jaw the cups of the corner teeth are elliptic and commence to disappear.

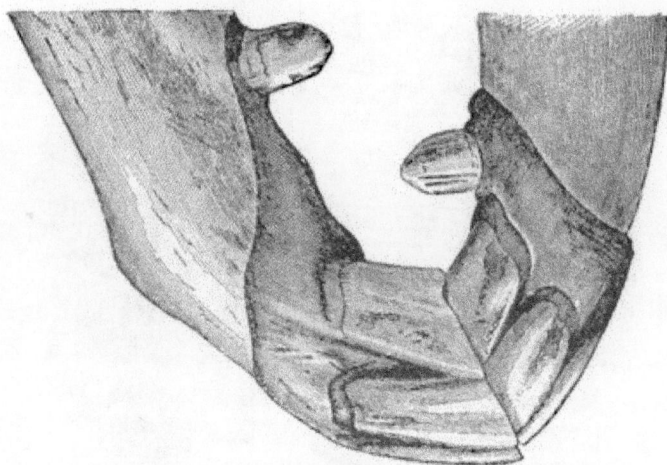

FIG. 101.

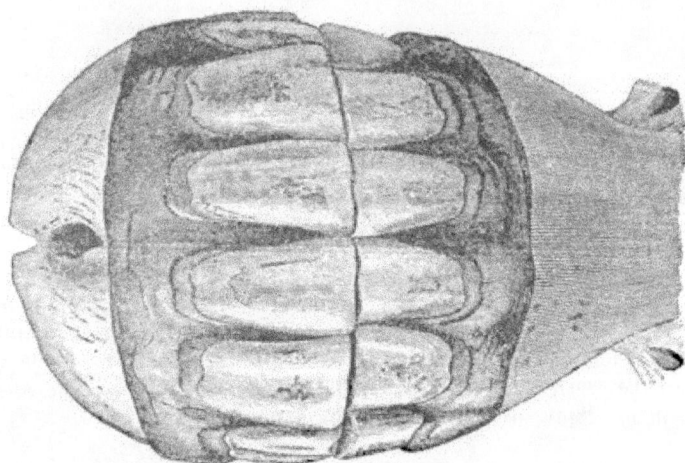

FIG. 105.

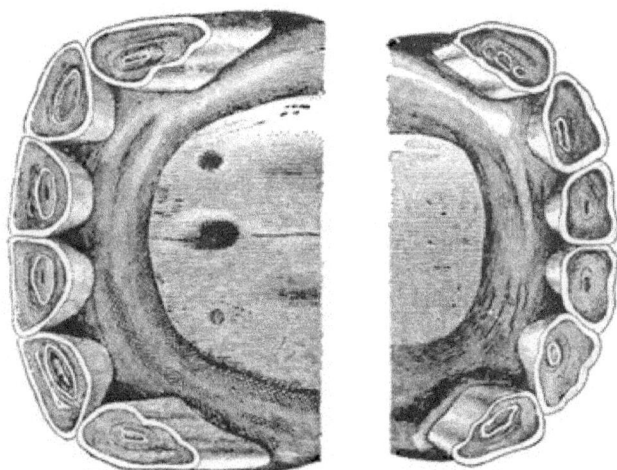

FIG. 106.

Twelve Years (Figs. 104, 105, 106).—The incidence of the jaws is still more oblique than at eleven years. In profile, the superior corner tooth is still more oblique; it has a notch on its posterior border, and it is separated farther from the intermediate. The inferior tables are round and nearly leveled, or only have traces of the cups. A yellow spot represents the dental star. The superior corner teeth are almost leveled. The incisive arches are narrower and less convex than at an earlier age.

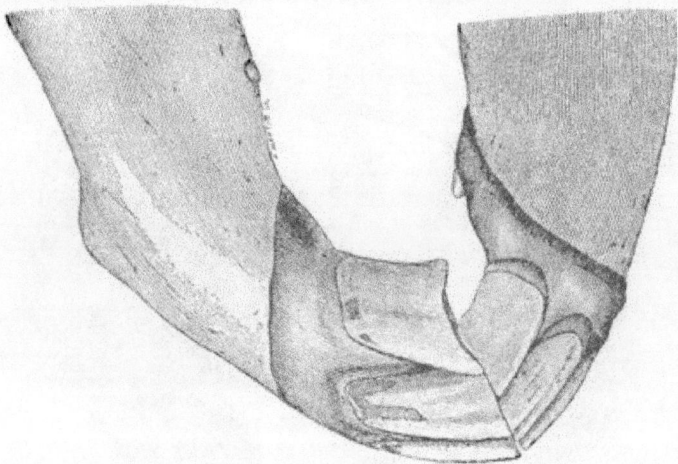

Fig. 107.

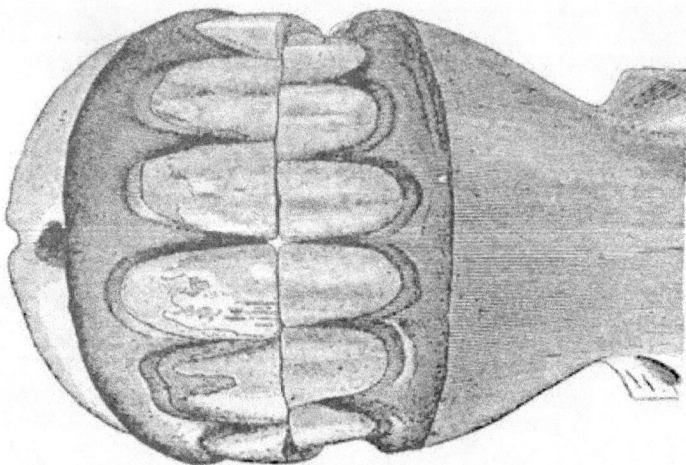

Fig. 108.

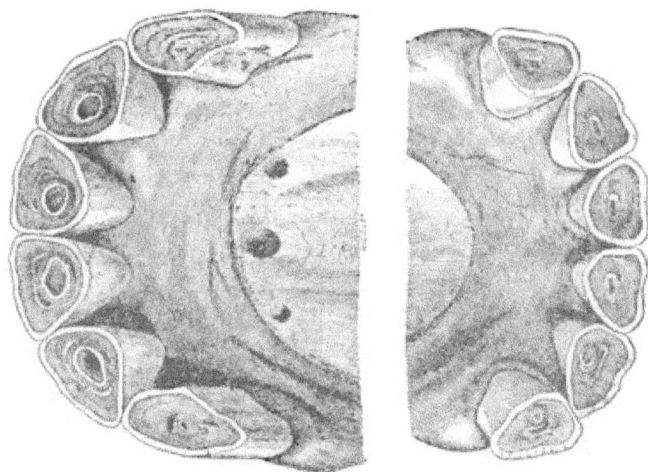

Fig. 109.

Thirteen Years (Figs. 107, 108, 109).—In front, the appearance is much the same as it is at twelve years. In profile, the notch of the superior corner tooth is larger; the inferior appears narrow, with its borders parallel. The plate, drawn from a mare's mouth, shows rudimentary tusk teeth. The inferior tables are round and generally leveled. The superior corner teeth are generally leveled. In the superior pinchers the cups are rounded.

FIG. 110.

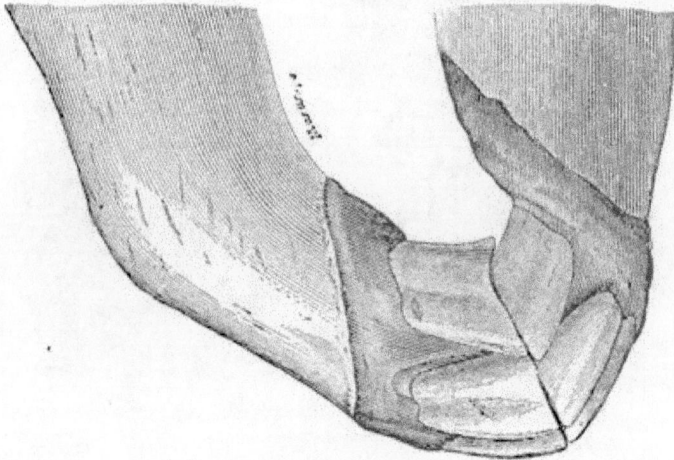

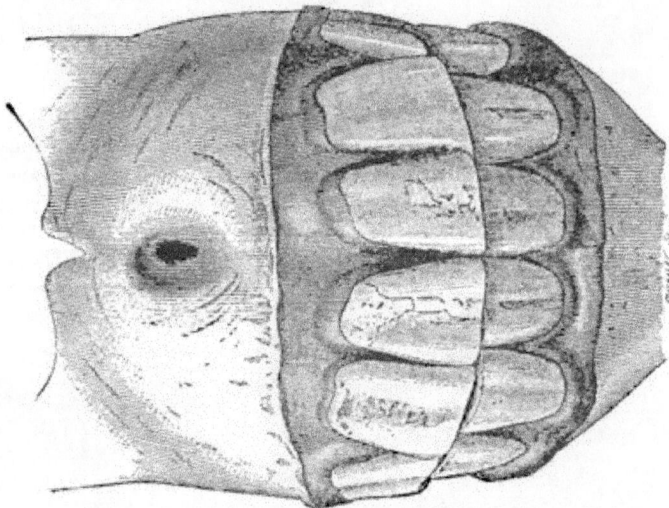

FIG. 111

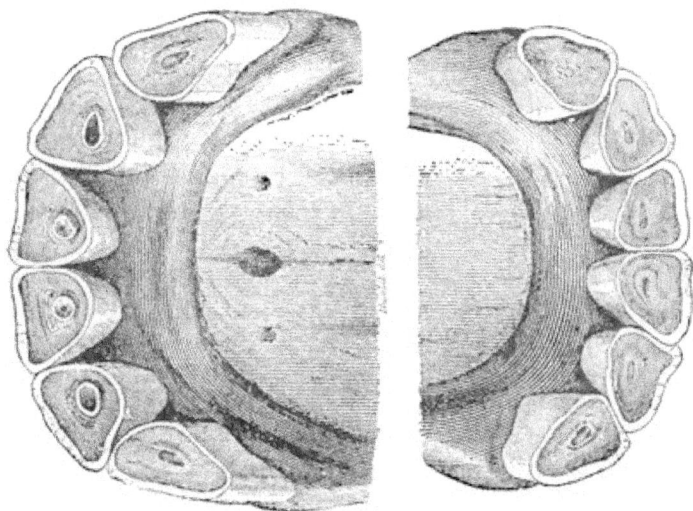

FIG. 112.

Fifteen Years (Figs. 110, 111, 112).—From in front the inferior teeth appear shorter than the superior, as the jaw has not been raised to the proper height; in profile, they are seen to be about the same length. The notch in the superior corner teeth continues. The inferior tables show, in their centre, a very decided, rounded dental star. The pinchers are nearly triangular in shape; the intermediate teeth begin to become so. The central enamel in the upper pinchers is much smaller than at thirteen years; the incisive arch is depressed in front and narrowed transversely. In this month the cups have disappeared from the upper corner teeth.

8

FIG. 113.

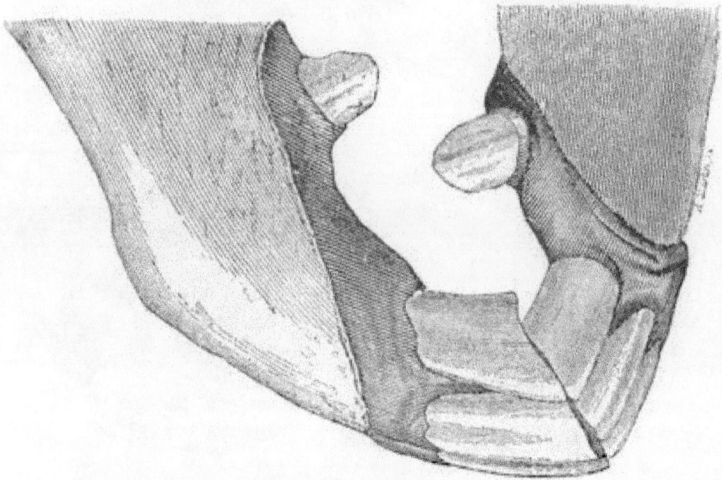

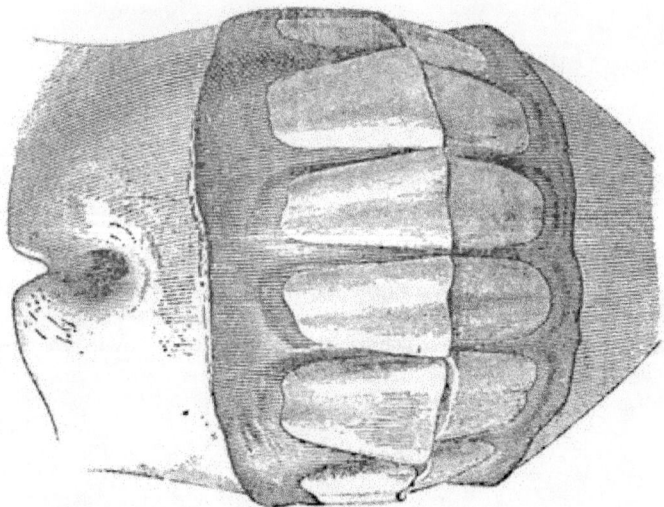

FIG. 114.

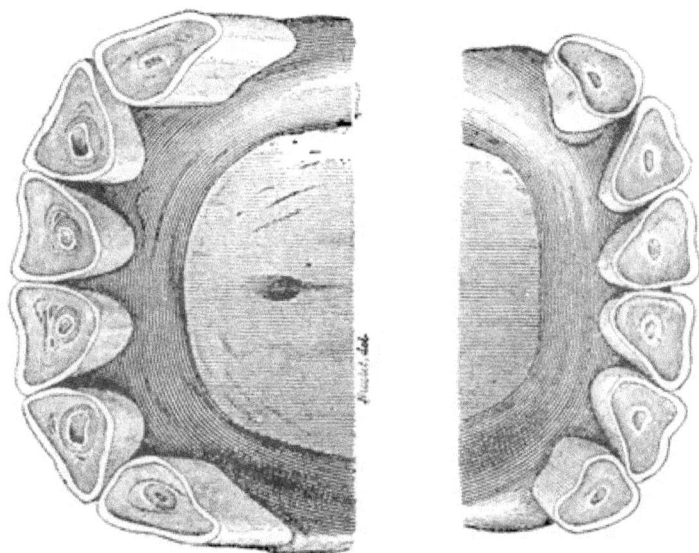

Fig. 115.

Seventeen Years (Figs. 113, 114, 115).—In front, the superior corner teeth incline toward the centre. The line of apposition of the teeth
is very oblique on the horizontal line. To see the inferior teeth distinctly, the head must be well lifted. The tables of the inferior teeth
are all triangular; the dental star is in the centre and distinctly round.
The inferior incisive arch is narrow, only slightly convex, and the teeth
appear more separated from each other than at any previous period; the
pinchers especially are separated from each other in the median line.
The tables of the superior teeth are triangular; the cups of the pinchers
have nearly disappeared.

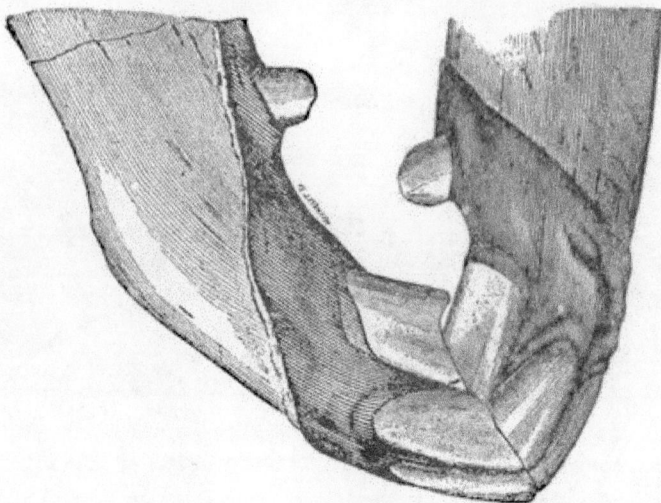

FIG. 116.

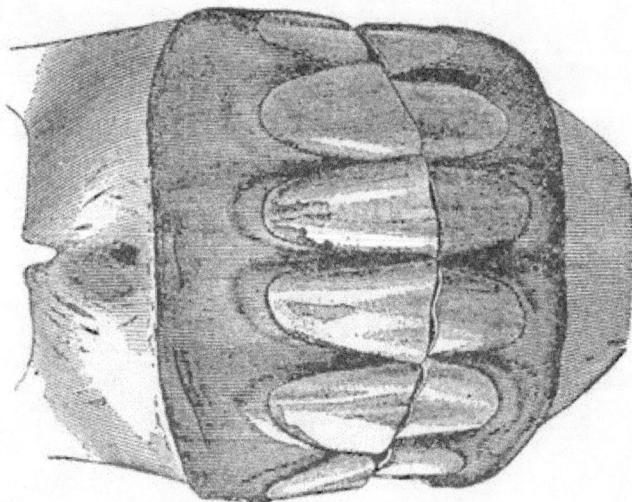

FIG. 117.

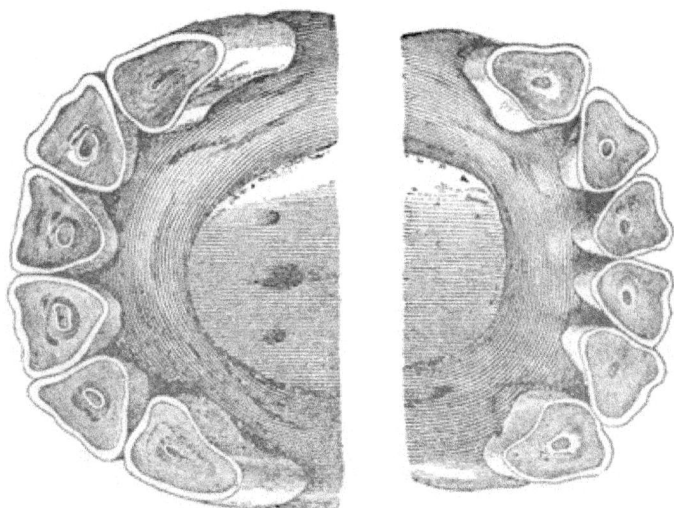

FIG. 118.

Nineteen Years (Figs. 116, 117, 118).—From in front the superior corner teeth point distinctly toward the median line; the intermediate teeth commence to incline in the same direction, showing a marked triangular space between them at the line of the gums. In profile, the ogive formed by the apposition of the jaws is more closed. The notch still remains in the superior corner teeth, but is less marked in this age from the increased horizontal position of the inferior corner teeth. The tables of the pinchers and the intermediate inferior teeth seem to converge at their posterior border, on account of their diverging in front; their antero-posterior diameters have become greater. The inferior corner teeth are always triangular in the upper jaw; the pincher teeth are usually leveled, although the cups sometimes remain in them for several years more.

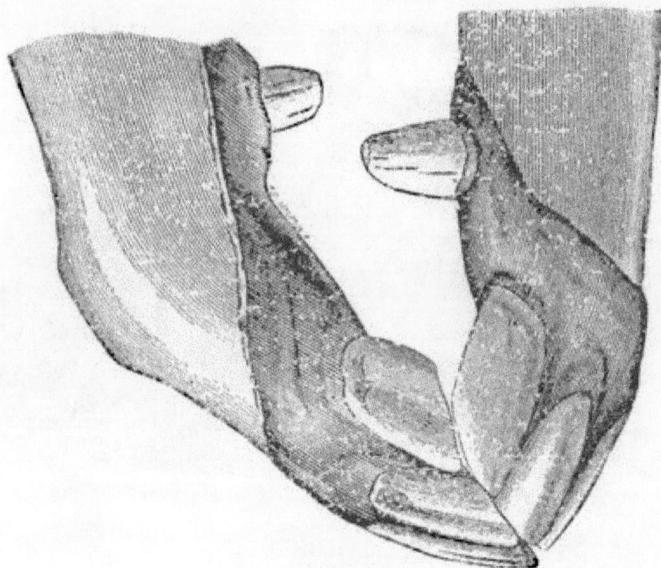

FIG. 119.

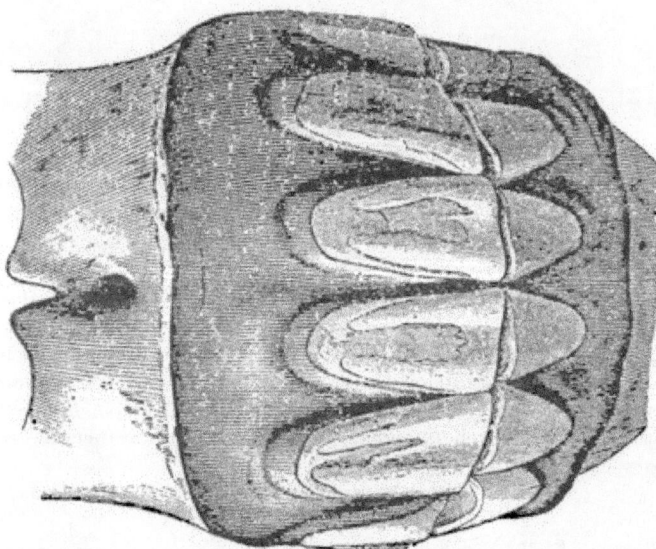

FIG. 120.

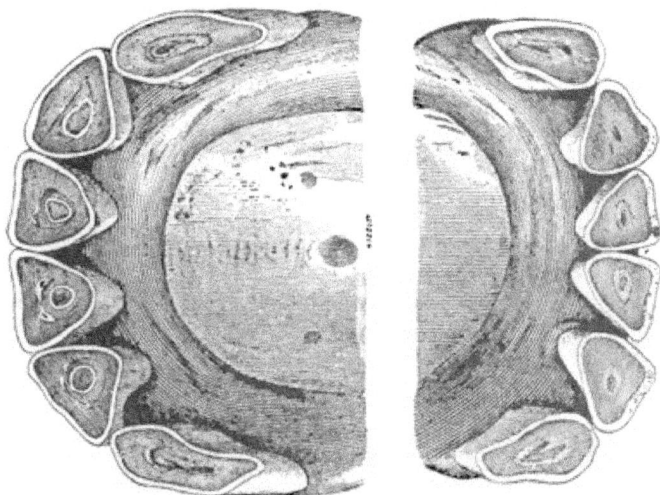

FIG. 121.

Twenty-one Years (Figs. 119, 120, 121).—The teeth become so horizontal that, looked at in front, the inferior ones are scarcely seen unless the head is well raised. The triangular interspaces at the base of the superior incisors increase in size and the convergence of the free ends becomes more marked. In profile, the jaws are seen diminished in size; the inferior corner teeth have become almost horizontal and have worn off the notch on the superior corners. The wearing surfaces of these teeth become elongated from in front to behind and have lost their triangular shape. The tables of the superior pinchers and intermediate teeth are elongated from in front to behind and distinctly triangular; they are generally leveled. The tables of the inferior teeth commence to flatten from side to side and are more distinctly separated.

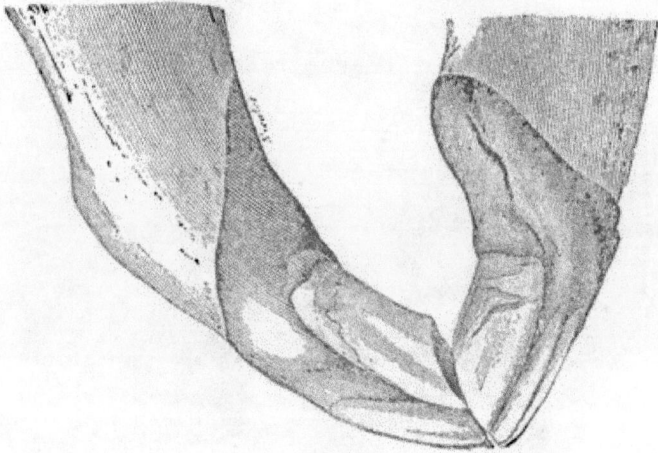

Fig. 122.

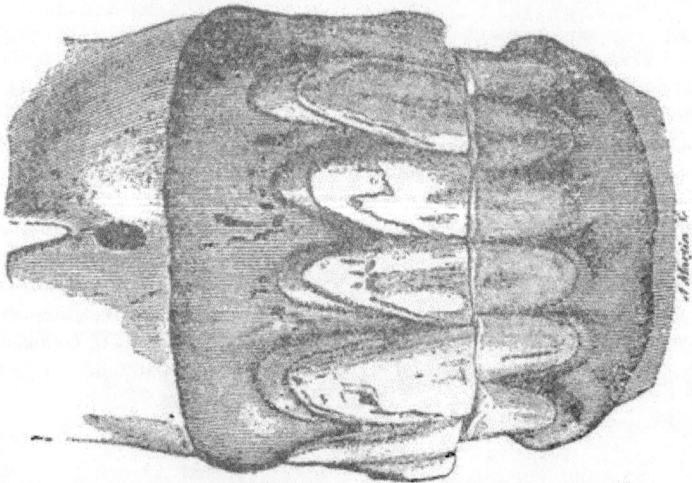

Fig. 123.

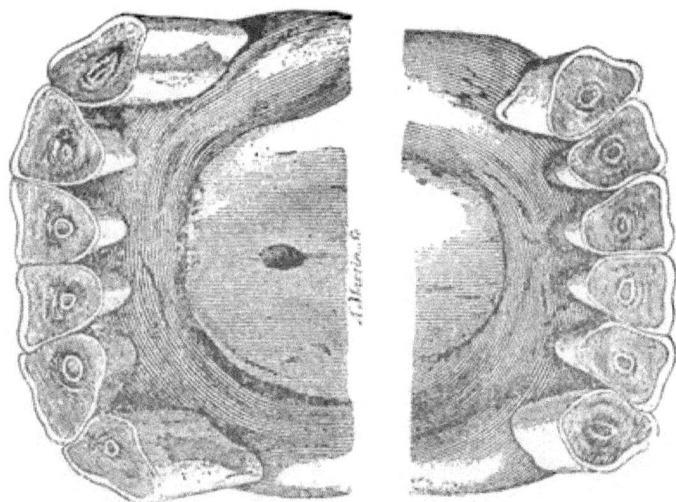

FIG. 124.

Thirty Years (Figs. 122, 123, 124).—This plate shows the character of extreme old age. In front, the superior incisive arch overlaps the inferior, which has become considerably narrower; the convergence of the corner and intermediate teeth becomes more marked. In profile, the inferior incisors are almost horizontal, especially the corner ones; the jaws are thinned and are wider apart from each other in the region of the bars. The tables of the inferior teeth are flattened from side to side (biangular). The peripheral enamel has almost disappeared from the posterior border of the teeth. In the upper jaw the tables are flattened from side to side and the enamel is nearly worn away. Sometimes in one or in both jaws the teeth have acquired an extreme length and are not leveled ; or, at other times they are worn away almost to the roots and level with the gums, and are surrounded by a large deposit of the radical cement which covers the dentine, while the enamel has entirely disappeared.

THE dental system offers a large number of very interesting and important irregularities. When these exist the horse is said to have an irregular mouth, or to be falsely marked. Some of the irregularities are without importance, while others may influence the wearing of the teeth to such a degree as to make the determination of age excessively difficult, except by a careful

 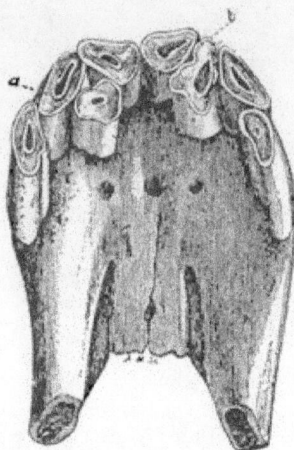

FIG. 125. FIG. 126.

study of all of the changes and a comparison of the indications furnished by the other parts of the body.

The irregularities can be classified as follows:—

1. Number.—augmentation and diminution.
2. Form of the incisors.
3. Uniting of two incisors.
4. The form of the cup; fissure of the dental cup.
5. Depth of the dental cup and size of its cavity (*bègue*).
6. Fault of length or excess of size of one of the jaws

(122)

(prognathism, brachygnathism); excess of length of the superior incisive arch.

7. Excess or fault of use.

8. Marks produced by cribbing.

9. Fraudulent alterations,—removal of the milk-teeth, bishoping. filing the corners.

IRREGULARITIES IN NUMBER—AUGMENTATION.

Incisors.—The most curious example of this anomaly is one which was noted for the first time by Lafosse, in 1772.

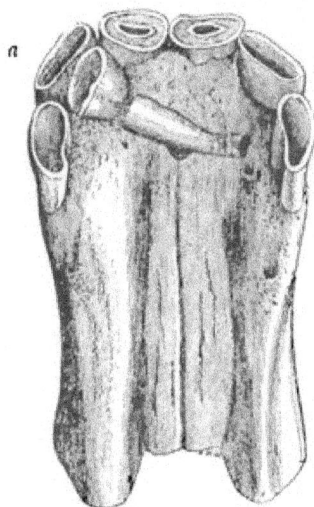

Fig. 125.

He states that he found horses with double rows of incisive teeth. This anomaly was again noted by Goubaux. in 1842. His case had two rows of incisors of second dentition in each jaw, making twenty-four in all.

Augmentation or duplication of the incisors of second dentition for one or two pairs of teeth is not rare. The accompanying plates give an excellent example. In Figure 125 there are two supernumerary pinchers (*a a*) and one supernumerary intermediate tooth (*b*). Figure 126 shows two supernumerary intermediate teeth (*a* and *b*). Figure 127 has an intermediate tooth

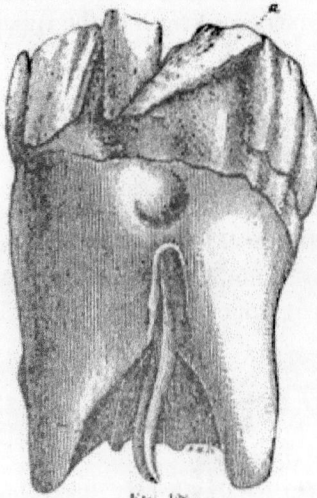

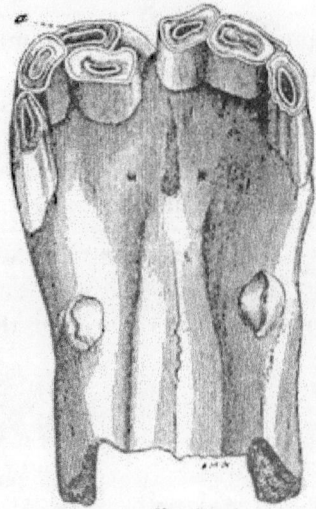

FIG. 128.   FIG. 128a.

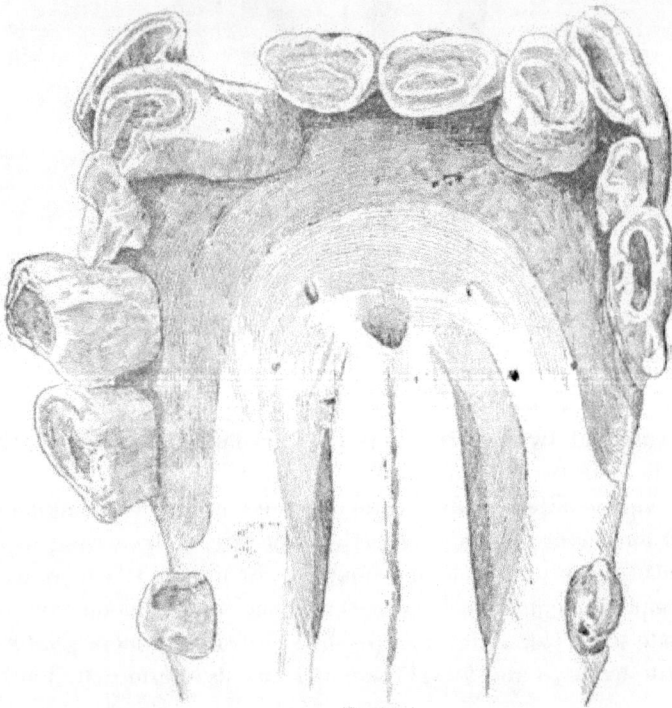

FIG. 129.

(*a*) directed transversely and held in place below the incisive foramen by a bony bridge.

Figures 128 and 128*a* show a pincher tooth similarly deflected.

All these teeth are of second dentition.

Supernumerary teeth of the lower jaw are less common than in the upper one; they sometimes, however, occur. The presence of supernumerary teeth rarely modifies the wearing of the others in the marking of age. Supernumerary teeth are usually firmly held in their alveolar cavities, and rarely alter the shape of the arch of the incisors except when milk-teeth are still remaining, with which they should not be confounded.

Figures 129 and 130 show an upper jaw from the collection of Dr. J. W. Gadsden, of Philadelphia, in which will be seen two supernumerary intermediate teeth, a right-hand supernumerary corner tooth, and the corner teeth of first dentition still remaining.

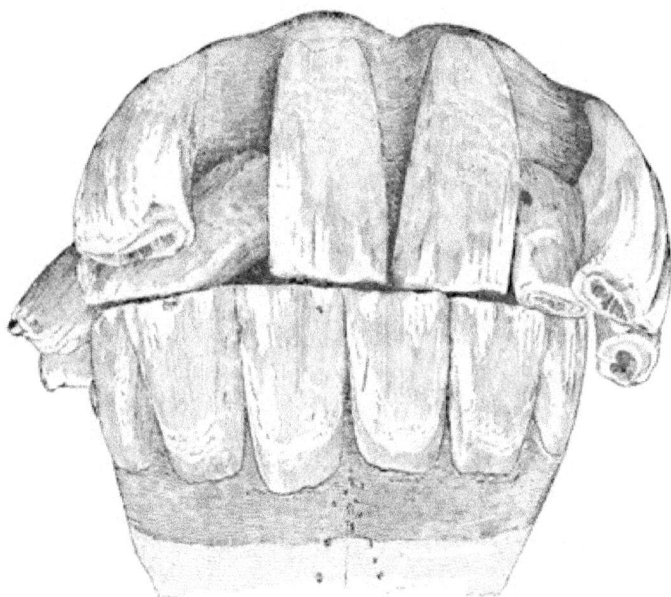

Fig. 130.

Tush Teeth.—Goubaux reports a case of an ass with a supernumerary tush tooth which seems unique (Fig. 131).

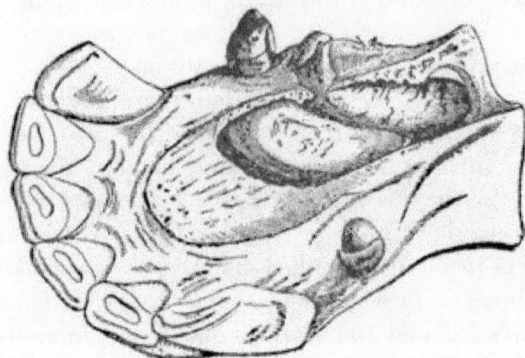

FIG. 131.

Molars.—Supernumerary molars have only been found in the upper jaw. They occur sometimes in line in the normal arch, at other times protrude to the outer side, in which case they may cause trouble by wounding the cheek. They have also been found under the zygomatic process at the base of the ear, where they may cause an abscess which opens on the exterior, and have been known to protrude into the cranium.

FIG. 132.

Figure 132 represents a sub-zygomatic molar, natural size, removed from an abscess under the ear of a three-month-old gelding, by Dr. W. J. Martin, of Kankakee, Kansas.

DIMINUTION.

Diminution in the number of the incisors is less frequent than the presence of supernumerary teeth. Diminution should not be confounded with cases of tardy eruption; nor with those cases of arrest of development, in which the teeth remain in their alveolar cavities; nor with cases of fracture or surgical injuries and loss of teeth. Diminution includes only the complete abortion of the dental follicles, and can sometimes only be determined by examination of the jaw after death.   Incisor, tush, or molar teeth may be absent.

The tush tooth is the one which is most frequently absent, —a condition, however, which is practically normal in mares.

IRREGULARITIES OF FORM.

In certain subjects the incisors of the lower jaw present, at the age of six years, a decided triangular form, such as is usually seen at fourteen to fifteen years of age.

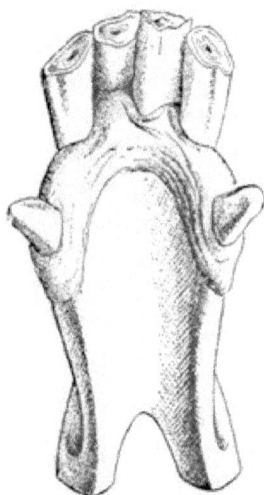

FIG. 133.
Absence of corner incisors.

FIG. 134.
Double cup of left intermediate tooth.

This triangularity is readily distinguished from that of the older age by the presence of the cup.

IRREGULARITIES BY UNITING OF TWO INCISORS.

This sort of irregularity has no special importance as to the determination of age, but is interesting as a curiosity.   Figure 134 shows the upper jaw of a horse with a double left-hand intermediate tooth, in which the two cups are perfectly distinct.

IRREGULARITIES IN FORM OF THE DENTAL CUP—FISSURE.

Fissure of the incisor teeth, by failure of the enamel on the posterior face to completely surround them, is rather more common in asses and mules than in horses. The fissure may represent only a small portion of the tooth, or it may constitute a complete division of the posterior surface, leaving the peripheral enamel in the form of a crescent (Fig. 136).

IRREGULARITIES IN DEPTH OF THE DENTAL CUPS.

Frequently the dental cup continues on the table of the teeth at a time when it should have disappeared. This occurs more frequently in very well bred horses in which the development of the teeth has been precocious, and in which the teeth have such a consistency and hardness that they do not wear away by use, as they do in more common bred horses.

EXCESS OF HARDNESS OF THE TEETH.

The nature of the food which the animal has eaten and the condition of health of the teeth are causes which tend to influence this irregularity. It is a condition rarely noticed before the age of seven or eight years.

The length of the teeth and the form of the dental tables are less to be considered in estimating excessive length of the teeth than is the form of the teeth and that of the cups.

EXCESS OF LENGTH OF THE DENTAL CUP.

Sometimes while the teeth have worn in the ordinary way, have diminished in length, and have undergone the usual changes in the shape of their tables, the cups continue in size and shape not corresponding with the other indications of age.

In these, an attentive examination of the tables of the incisors with reference to their shape; the size of the dental star, its situation; the condition of the upper incisors; the direction of the teeth, their length, color, etc., will rectify the faulty indications furnished by the cups.

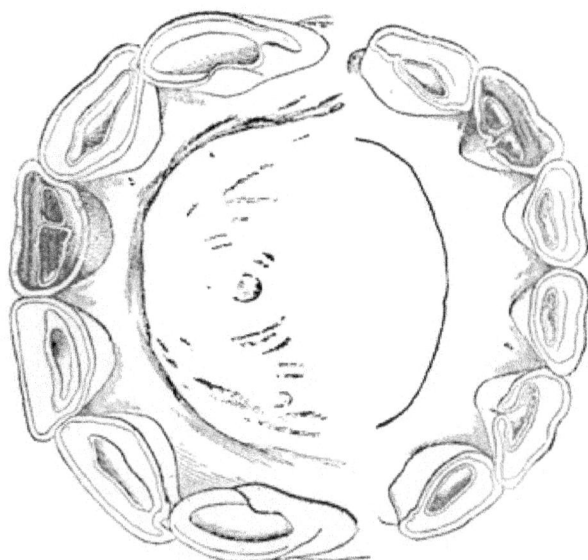

FIG. 135.
Double cup of left upper pincher and of left inferior intermediate tooth.

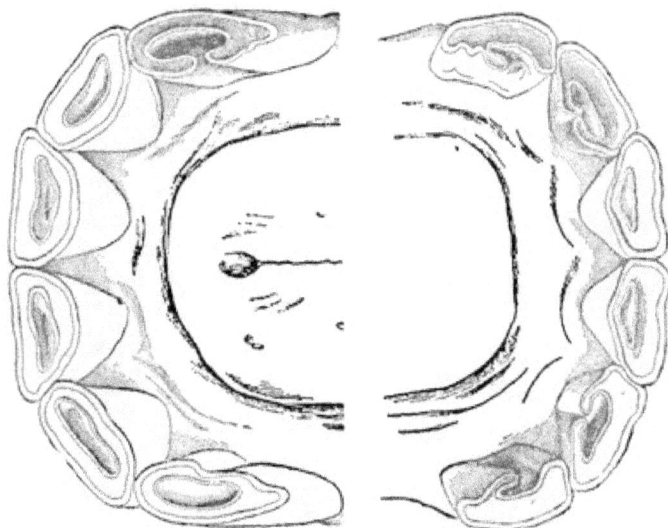

FIG. 136.
Fissure of the cups of the inferior intermediate and corner teeth and of the left
superior corner.

9

Fig. 137.

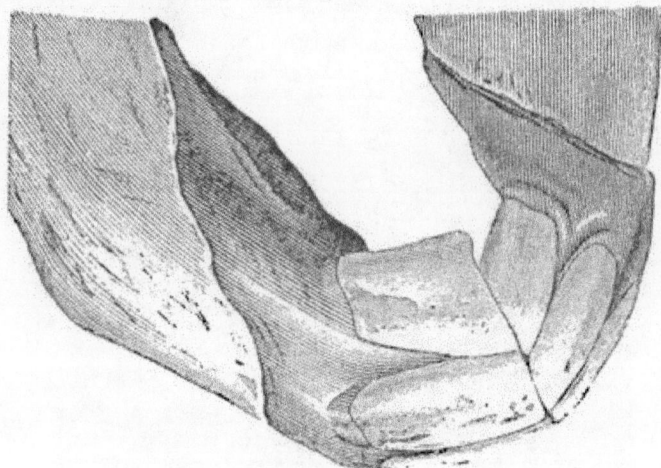

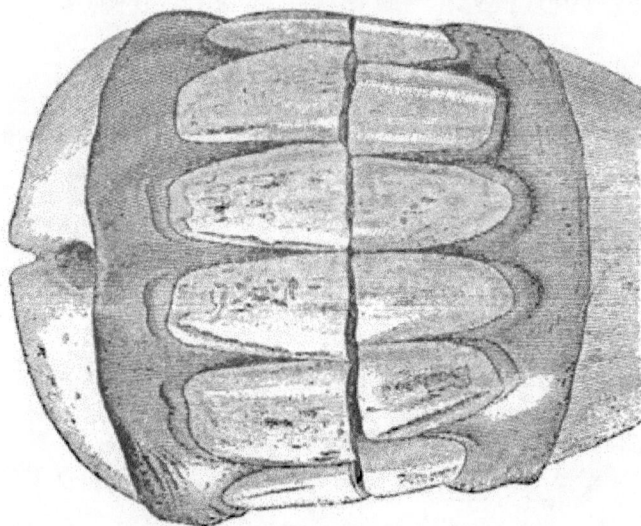

Fig. 138.

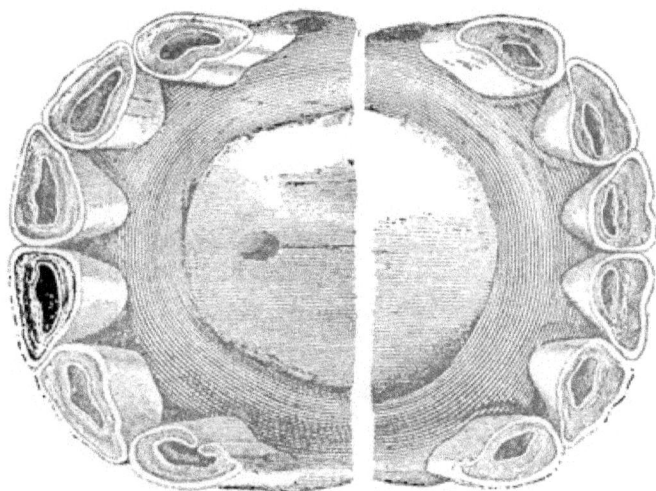

FIG. 139.

Nine Years (Figs. 137, 138, 139).—These plates are from the teeth of a horse nine years of age; the dental cups are of a size and shape representing a younger age. In profile, the incidence of the teeth is seen to be that of a horse five or six years of age; the inferior tables are round; the central enamel incloses small cups close to the posterior border of the teeth. The corner teeth are leveled and rounded; the dental star is near the middle of the teeth. These characters, added to the freshness of the corner teeth, the obliquity of the inferior incisors, the notch on the upper corners, and the general condition of the subject, are sufficient to rectify any error indicated by the first inspection.

FIG. 140.

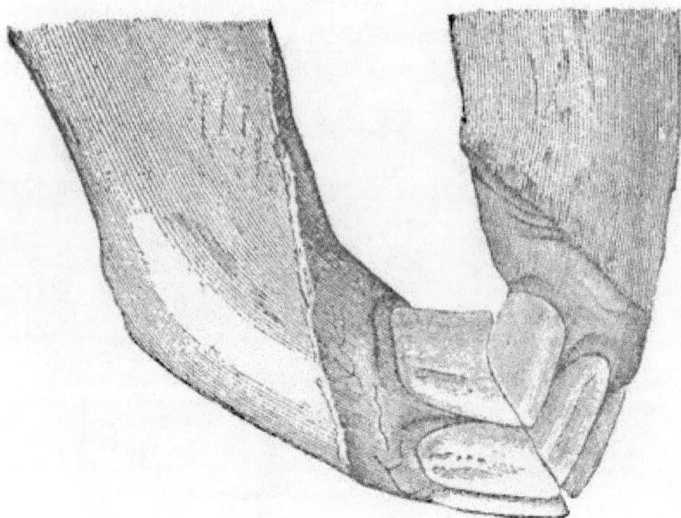

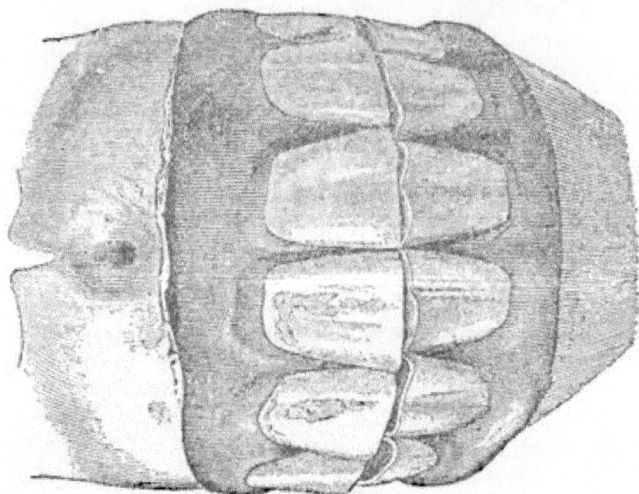

FIG. 141.

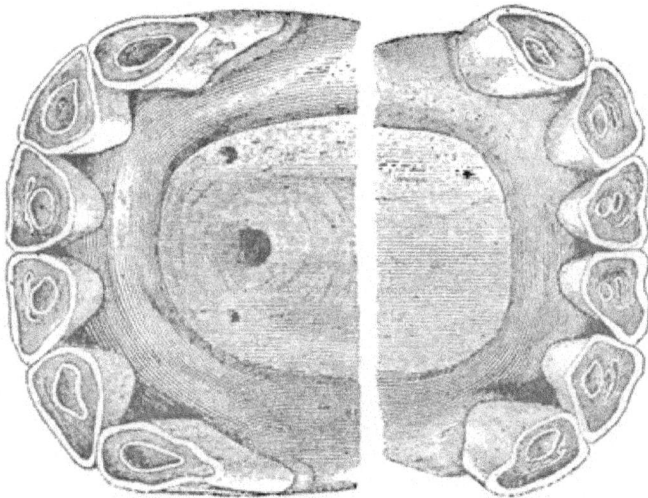

Fig. 142.

Fourteen Years (Figs. 140, 141, 142).—This figure is taken from the mouth of a horse fourteen years of age. The leveling of the teeth would only indicate about ten years of age, but a close examination of the tables shows (1) that the pinchers are nearly triangular; (2) the intermediate teeth commence to become so; (3) the dental star is perfectly distinct, narrow, and rounded; (4) the inferior incisive arch is depressed in its centre. Seen from in front, the teeth show no cement, and the inferior teeth are so horizontal as to be hidden unless the head is raised. In profile, the inferior corner tooth is no more oblique than that of a horse ten years of age, is narrow in front to behind, and has about the same diameter its entire length. The superior corner tooth is notched. The incidence of the incisive arches is acute. All of these characters indicate that it is older than would be supposed from the first observation of the teeth.

IRREGULARITIES FROM FAULT OF LENGTH OR EXCESS OF SIZE OF ONE OF THE JAWS.

Deficiency in length of either of the jaws is rare in the horse, but sometimes occurs. The simplest form is where there is a slight projection of the lower jaw (prognathism in man); in this there is excessive wearing of the lower incisors, which complicates the determination of age (Fig. 143). A more serious abnormality is deficiency in length in the lower jaw (brachygnathism in man), which is followed by deformity of the intermaxillary bones; these curve down and interfere greatly with the prehension of food, especially at pasture.

In prognathism the lower table may project for a variable distance beyond the upper jaw, but it is rare that it extends beyond to any marked degree, and, while it causes the teeth to wear away rapidly, there is not often any alteration in the incidence of the jaws. In brachygnathism the difference in length of the two jaws may be very great, in some cases so great that the upper jaw completely overreaches the lower one, and by bending of the intermaxillary bones the upper teeth drop over the anterior face of the lower teeth, while the latter point their tables directly into the hard palate; in other cases the posterior face of the upper incisors wears against the anterior face of the lower incisors until both are beveled into sharp wedges (Fig. 144).

EXCESS OF WIDTH OF THE UPPER INCISIVE ARCH.

This condition is normal in old horses, without inconveniencing them. It sometimes occurs in young horses, in which the irregular wearing against the upper teeth produces marked notching and sometimes interferes with the prehension of food; in one case, in a well-bred Hambletonian mare, in which the excess of breadth of the upper incisive arch was due to the presence of supernumerary corner teeth, giving the arch eight teeth in all, the deformity was very marked.

IRREGULARITIES BY EXCESS OR FAULT OF USE.

As a general thing the incisive teeth have the same length in their free portion, although the whole tooth is constantly

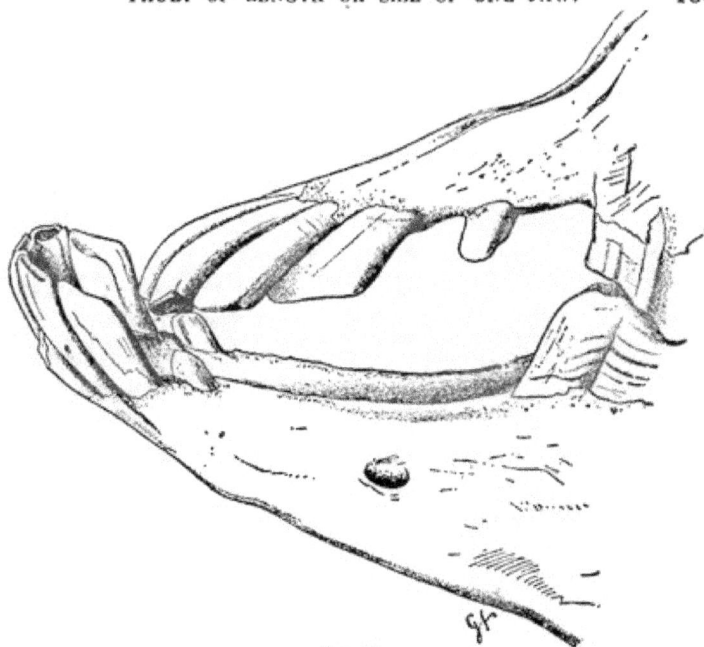

FIG. 143.
Superior brachygnathism.

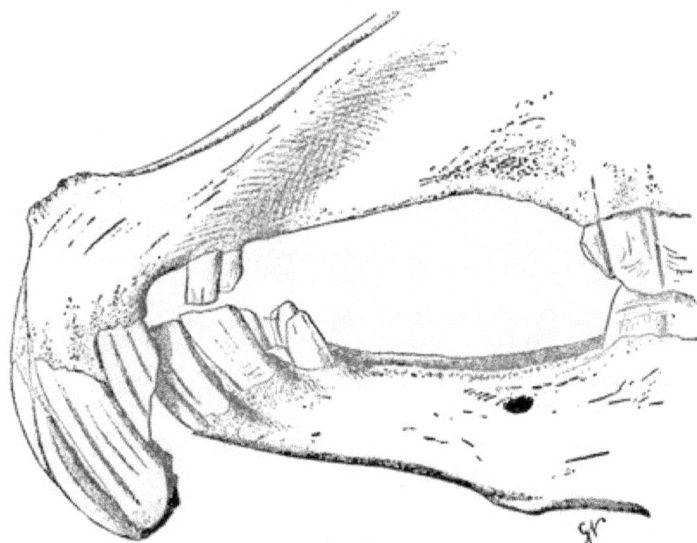

FIG. 144.
Inferior brachygnathism.

shortening each year, from the wearing away of its table; this even length is maintained by the roots being constantly forced from the alveolar cavities and becoming a part of the free portion; transverse lines, filed at definite points on the anterior surface of the incisors, will be seen to approach the table and disappear, one after another. Experiments by Pessina showed that in common horses the annual wearing is about 4½ millimetres, while in thoroughbreds it is only about 3 millimetres; experiments by Bouley verified those of Pessina.

Girard, after numerous investigations, determined that the free portions of the incisors, from the gum to the table, average 15 millimetres. This is subject to some variation, but normal pinchers rarely vary from 18 millimetres in length, intermediate teeth beyond 15 millimetres, while corner teeth are about 13 millimetres. Variation in the length of the incisors may be from excess of length or from deficiency of length.

EXCESS OF LENGTH.

This anomaly occurs in several ways, either occurring in both jaws, or in the teeth of the upper jaw alone, or at times only in certain teeth of one or the other jaw.

EXCESS OF LENGTH OF BOTH JAWS.

When the teeth of both jaws are too long, there is a general tendency for them to become parallel; that is, to approach a horizontal direction; but their free extremities are diverged like the ribs of a fan, and the wearing surface, or table, shows the character of the age of the horse; flattened in front to behind, they do not tend to take an oval form; the central enamel occupies a large portion of the dental table. There is often a little external cavity in the inferior corner teeth; the excess of length of the crown of the tooth is not in proportion to that of the root; the teeth are less solidly fixed in their alveolar cavities, and are subject to fracture from moderate violence, on account of the great leverage of the lengthened crown. These teeth on their table apparently show an age which is sometimes very deceptive.

To determine the exact age, it is necessary to shorten, by the imagination, the elongated teeth; determine by a close examination what would be the form of the tables of the teeth were they cut off to their proper length, what would be the form and position of the cups, and dental stars in such a shortened tooth, and give to the animal the age which the alterations indicate.

Cutting off incisors of great length in a young horse need not be considered a fraudulent act on the part of the dealer, for, while shortening of the long teeth of old age is a trick to deceive the ignorant, who associate long teeth with great age, the shortening of the teeth of a young horse gives a table indicating a greater age, and will not deceive the expert, who should always examine any horse on purchase.

EXCESS OF LENGTH OF THE INCISORS OF THE UPPER JAW.

This anomaly constitutes what is commonly known as *parrot-mouth* (Figs. 145 and 146), on account of the analogy in the appearance of the upper jaw to the corresponding beak of this bird. The upper teeth may acquire a length of 2½ inches, and are frequently very much curved forward and downward, while their posterior faces are worn away to a sharp bevel by their contact with the inferior incisors. These latter are frequently shorter than normal, and the parrot's beak is formed by the pincher and intermediate teeth, with only the internal border of the corner teeth, while the rest of the latter is worn into a deep notch. These deformities are frequently much greater on one side of the jaw than on the other.

In some horses five years old the upper jaw projects a line or two in front of the inferior jaw, whilst the posterior surface of the teeth in both jaws corresponds, which produces excessive wearing of the posterior part of the upper incisors, leaving a little line in front of their table, which predisposes to the formation of parrot-mouth. Usually, however, parrot-mouth is only seen in very old horses. It interferes more or less, according to its development, in the prehension of food, especially with oats, as the projection of the teeth interferes somewhat with the move-

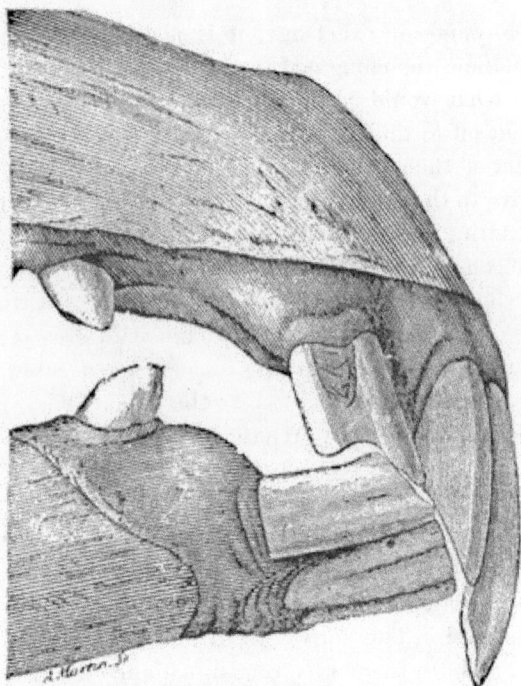

FIG. 145.

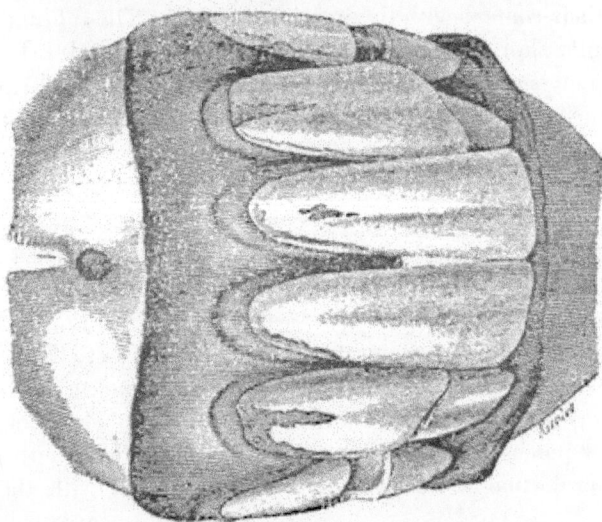

FIG. 146.

ments of the lips; the elongated teeth also render less the maxi-
mum extent to which the incisors can open than in a normal
mouth; the latter can usually open 2½ inches, or a little more,
while in certain parrot-mouths the space between the open incisors is only 1½ to 1½ inches. The beveled and distorted tables
of the incisors in a parrot-mouth render the determination of
age, for these, practically impossible. The judgment of age
may be based upon the relative characters, as to the inclination
of the teeth, their color, their size, and what one supposes would
be their tables were they cut to their proper length. It
frequently happens that a parrot-mouth so interferes with mastication, on the part of the animal, that it becomes necessary
to resort to operative measures with the saw and file to give
the animal a normal mouth. This operation not only relieves
the animal, but restores it to the appearance of its approximate
age, and does not enter into the category of fraudulent measures.

EXCESS OF LENGTH OF THE INCISORS OF THE LOWER JAW.

A condition which might be called a reversed parrot-
mouth is sometimes seen, which seems to be caused by a slight

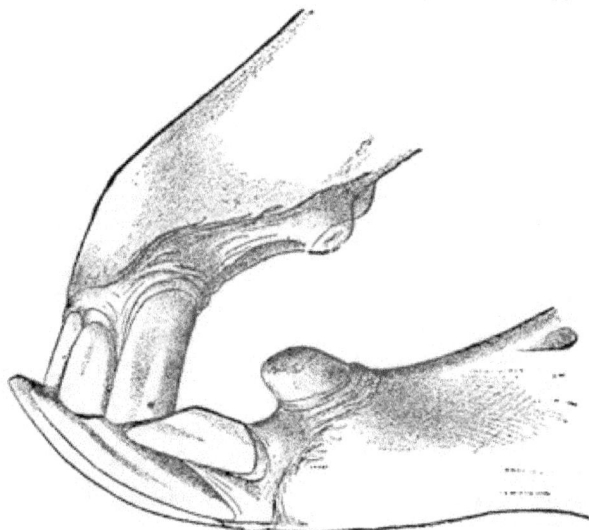

FIG. 147.

prognathism of the lower jaw, or a deficiency of the teeth of the upper jaw. It is usually limited to the pinchers and intermediate teeth (Fig. 147). This condition is more frequent in mules than in horses, and produces great inconvenience in seizing food, frequently requiring operative interference.

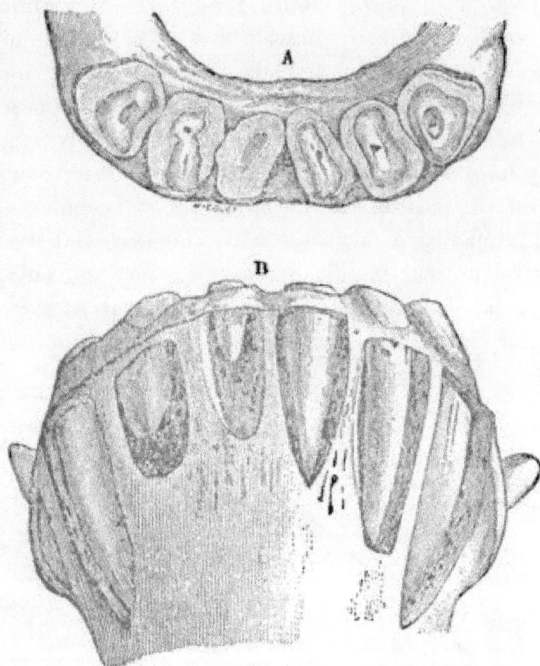

FIG. 148.
Deficiency of length of incisors of the lower jaw. A, dental tables. B, roots shown in their alveolar cavities.

EXCESSIVE LENGTH OF SINGLE TEETH OF THE JAW.

Mouths occasionally are seen in which only one, two, or a small number of teeth are distorted; this occurs at times on one side only, or it occurs frequently that the teeth on one side are very long and those on the opposite side very short, corresponding with an inverse condition in the other jaw. In cases like this it should always be the rule, as is necessary in all horses to determine age, to inspect them from both sides. In horses of

seven or eight years old, a slight increase of wearing on one side
of the jaw will frequently make, in the table-mark, a deviation
of a year or two in the appearance of the animal's age, which is
readily corrected by a further examination from the other side.

DEFICIENCY OF LENGTH OF THE FREE PORTION OF THE INCISORS.

Deficiency of length is seen only in very old horses; the
teeth are always in apposition, except in the case of cribbing
horses, when they may be so worn off as not to reach those
of the other jaw. The diminution in the length may be so
great that nothing but a mere stub remains. This condition is
frequently accompanied by *cementoma*, or tumors thrown out by
the irritated alveolar periosteum (Fig 148).

MOLAR TEETH.

As in the case of the incisors, the irregularities of the molar
teeth are due to excess and deficiency of length. This may
occur on one or both sides of the jaw, on all of the tables of an
arch, or only on certain teeth.

DEFICIENCY OF LENGTH OF THE CROWN.

Inferior Jaw.—The abnormal use of the molars rarely extends to all of them in an arch; sometimes those in the centre
of the arch are the most worn (Fig. 149), giving it a concavity,
in which the corresponding elongated teeth of the upper jaw fit;
or sometimes they are worn most at the extremities of the arch,
with an inverse condition of the upper jaw (Fig. 150); in molars
which have been excessively worn the crown may have disappeared, leaving the roots as distinct and apparently supernumerary teeth (Fig. 149). In this case, as was noticed in the
study of the structure of teeth, when they are worn down, they
produce an irritation of their alveolar cavities which causes an
excessive deposit of callous bone or cement, which unites them
together and aids them in their function, but sometimes causes
annoying bony tumors; however, the new function is not as
complete as the original one, as the hard enamel has disappeared

and the wearing surface is much softer; the mastication is not complete, and the animal shows the effect of a lessened nutrition.

Superior Jaw.—What has been said of the lower jaw applies to the upper, but the changes in them are less frequent.

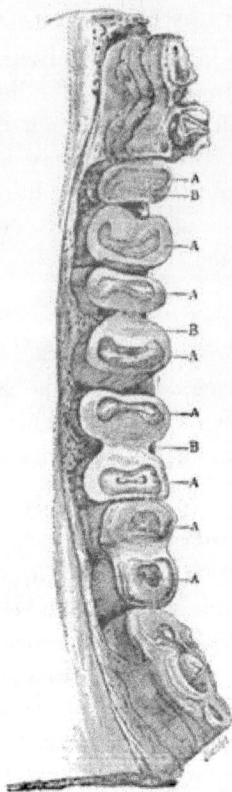

Fig. 149.
Lower molar arch of a very old horse.

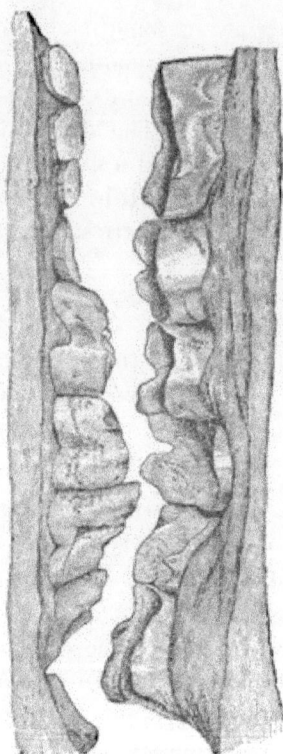

Fig. 150.
Molar arches (right) of a very old horse.
Lower jaw to the left of figure.

The separation of the roots in the upper jaw is less common, as these teeth are longer and contain a greater amount of enamel.

EXCESS OF LENGTH OF THE CROWNS.

Excess of length of the crowns or portions of them is of rather common occurrence. As the upper molars are broader as

well as longer than the lower ones, the lower jaw is obliged to
have a lateral movement in order to grind against the whole
table of the former.   When the horse is fed under the ordinary
conditions of nature, finding coarse weeds, twigs, hard vegetable
matters and resisting substances, as dirt, etc., mixed with its
food, this lateral movement is rendered complete by the slowness
and difficulty of mastication, and the teeth are kept worn evenly.
When, however, the animals are given selected fine hay, cleaned
oats and slops and mashes, easy of mastication, they use the
jaws like a chopping machine, and all parts of the grinding
surfaces are not brought into contact.   Again, when horses,
like cab-horses, doctors' hacks, etc., are fed at irregular times
and in a hurry, they masticate incompletely, give the jaws but
little lateral movement, and the narrower, lower jaws wear only
against the internal portion of the tables of the upper teeth.
When from any cause, such as caries of a tooth, cementoma,
disease of any kind, or deformity, the animal chews more on one
side than the other, the normal bevel of the teeth throws the
wearing surfaces from their natural position, and, while some
portions are worn faster, others do not receive the proper amount
of use.   When one or more teeth have been partially destroyed
or lost, the opposing teeth, finding lessened resistance, become
excessive in length and frequently irregular in shape.   The
irregularity of shape from any of the above causes is usually in
the form of an increased bevel.   Only the internal surface of
the upper molars is worn, causing the formation of a little ledge
along their outer borders.   This acts as a check to lateral
motion of the lower jaw, soon prevents it altogether, and cause
and effect promptly react on each other.   The external borders
of the upper jaw become long, thin, and sharp, and their irregular, ragged edges cut the cheeks and serve as lodging places for
balls of food, which may decompose, causing further irritation
to the mucous membranes.   The internal borders of the lower
teeth may become very long and stand up like so many arrow-
points, cutting the tongue and even penetrating the hard palate.
The pressure on the border of the teeth worn close to their

roots and the lateral pressure due to the leverage of the crown
on its root in the alveolar cavity produce further irritation and
prevent proper mastication.   The pain on one side frequently
confines the grinding entirely to the other.   The loss of the
lateral motion also prevents the slight fore-and-back play of the
arches of the teeth, and soon the anterior edge of the first molar
above and the posterior edge of the last molar below grow into
points, which may be very annoying to the animal; removal of
the latter is troublesome to the surgeon.   Disease of a molar in one

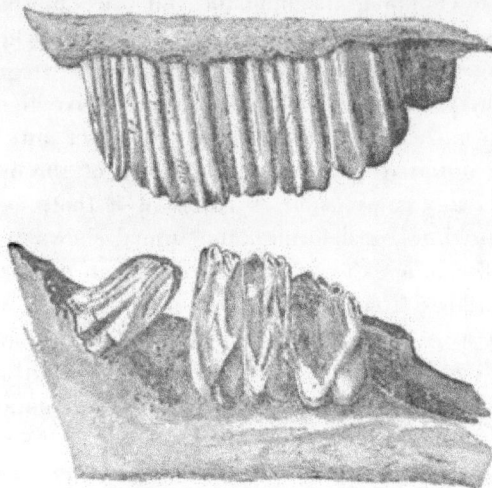

Fig. 151.
Irregular wearing of upper and lower molars.

jaw, or its absence from fracture or otherwise, is soon followed
by complication in the opposing teeth of the other jaw.   Fig.
152 shows a fourth molar in the upper jaw with a cavity which
was the size of a hen's egg, into which points the fourth inferior
molar, which, from want of resistance, has become elongated.
The presence of a dental cyst probably explains the enlargement
of the upper tooth and also the softness of its texture which
allowed it to be worn hollow.
     Irregularity and deformity of the molars are the causes of

much constitutional trouble; the interference with the mechanical action of the jaws and the soreness produced in them and in the cheeks and tongue render trituration of the food and mixture with saliva incomplete; the unprepared food is not digested and assimilated properly, causing a defective nutrition; the animal falls away in flesh, becomes hide-bound, and may have attacks of indigestion. The non-assimilation of the food causes indigestions and atony of the digestive tract and predisposes to intestinal calculi. The molars of all stable-fed animals should

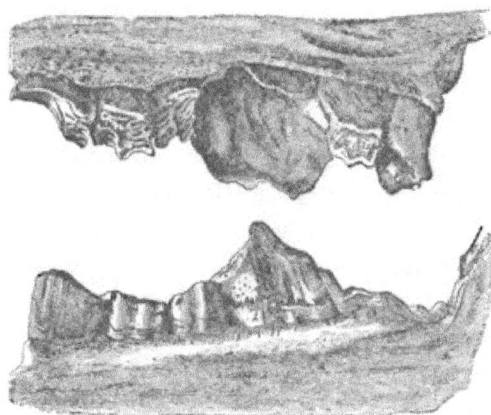

Fig. 72.

Hypertrophy of the upper fourth molar (right) with cavity for elongated corresponding molar.

be looked to once or twice a year, including those of race-colts which are grain-fed from weaning. The inspection of the molars should constitute a part of examination for soundness.

ACCIDENTAL IRREGULARITIES.

Accidents from various causes, injuries, wounds, fractures of the jaw-bone, with or without loss of the teeth, and abnormal growths, may so interfere with mastication on one side of the mouth that they cause an excessive wearing of the teeth on the other side.

Vicious, nervous, and irritable horses frequently bite at

10

their mangers, chains, wagon-pole, or other foreign bodies, and chip off the edges of the incisors, or they may even wear them to a considerable degree, so as to greatly change their form and tables and interfere more or less with the characters from which we judge the age. In many of these cases, however, the cause is evident, and the unilateral wearing, or the rough, irregular loss of substance, can be replaced by the imagination, and the accurate age determined.

Cribbing is recognized as of two kinds: 1st, that of the wind-sucker who pursues the habit, nose in air, and consequently produces no abnormal wearing of the teeth; and 2d, the cribber, who requires some foreign body between its teeth and wears them at the point of prehension. For the former, M. Goubaux proposed, in 1866, the name of "aëropinie."

According to the manner of cribbing and the character of the object which the horse chooses for support of the teeth, the wearing of the latter may be much varied. The object seized may be the feed-box, the rail of the manger, or a part of the stall, which may be horizontal or may be vertical; it may be the end of a poll or shaft, a chain, hitching-strap, or part of another horse's harness; in one case of the writer's, the horse would only crib on a small piece of wood when hung loose on a cord. Rare cases crib by seizing their own legs. Sometimes the support is only taken with the lips or the tuft of the chin, and in these there is no wearing of the teeth.

According to the size of the object, or the position assumed by the horse in cribbing, the contact of the teeth may be only by the anterior borders, only by the posterior borders, or by both; it may be by one jaw or by both; it may be by a number of teeth, or only by one or two teeth. In whatever manner the cribbing is done, but little force is used, and the worn surface of the teeth is smooth, even, and polished, so that it is readily distinguished from the roughened edges of teeth worn by vicious or nervous biting. When the support of the teeth has been taken, the animal makes an effort of deglutition, which is followed by a peculiar "cluck" sound. In examining an animal, even where the age is readily recognized, care should be taken to open the jaws completely, so as to inspect all surfaces of the teeth. While cribbing marks, when on the anterior face of the teeth, are apparent on superficial examination, those on the posterior face are often hidden by the foam and saliva, unless care is taken to wipe the latter away. A slight bevel worn on the anterior surface is evident, with the teeth closed, from the separation of the enamel and the presence of a yellow line, made by the exposed dentine, while a considerable bevel might be overlooked when its surface is a continuation of the yellow dentine of the table and is looked at from in front.

It is often of the greatest importance to determine if the irregular areas of use on a horse's teeth are due to accident, or are the result of cribbing. Undoubtedly, some cases of cribbing find their origin in the bad habit of cleaning a horse in its stall, where it can bite at and seize the manger, etc. In these cases the edges of the teeth are rough and jagged, and it is only after the habit becomes confirmed that the worn parts become smooth and polished and evenly used in their irregularities. In other cases, where attempts have been made to cure the horse of the habit, by fastening it so that it cannot reach the object on which it cribs, it changes its mode of cribbing and may wear various parts of the incisors, and then the teeth, losing the support of the enamel in front and behind, become very irregular and broken.

When the animal cribs by pressing the incisors against some foreign body, the wearing takes place on the anterior face of the teeth (Fig. 153, A, B, C, D).

When the object is seized in the mouth it may wear only the posterior faces (Fig. 154, A, B, C); or, if a thin board or similar body is held, it may wear the anterior borders of one jaw and the posterior of the other (Fig. 154, D, and Fig. 155, A).

If a thin object is seized evenly between the jaws, the tables of the upper or the lower or of both sets of incisors may be worn. In the latter case the age must be determined by the

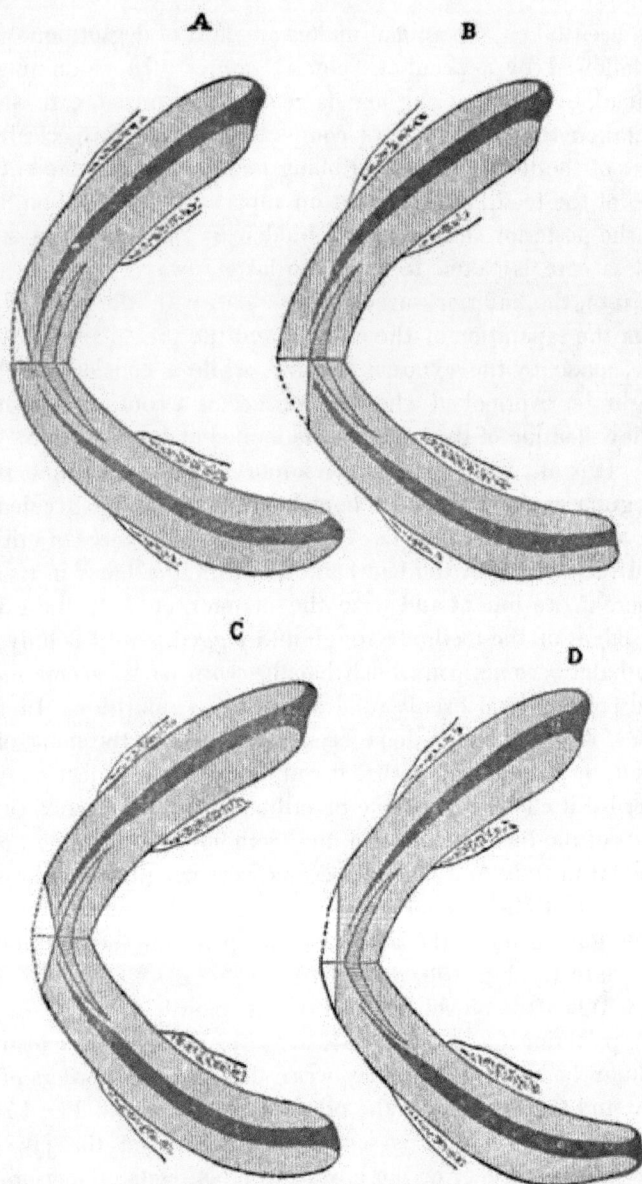

FIG. 153.
Crib-marks, by pressure of the teeth against foreign bodies.

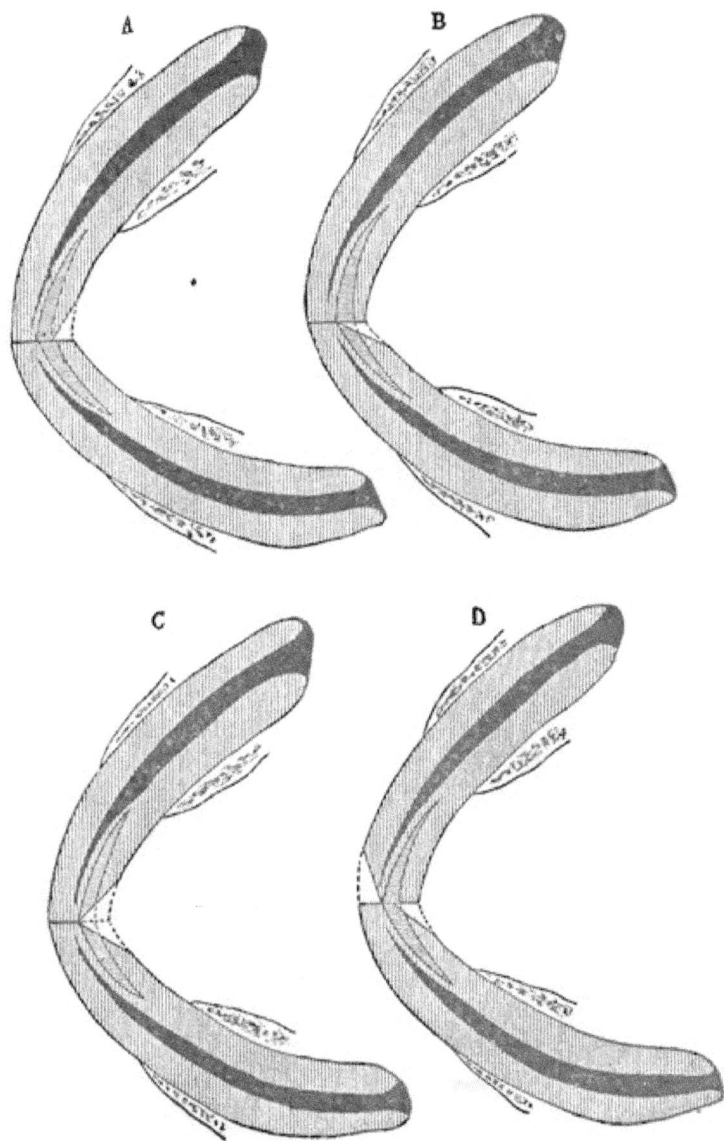

FIG. 154.
Crib-marks, by seizing foreign bodies inside of the jaws.

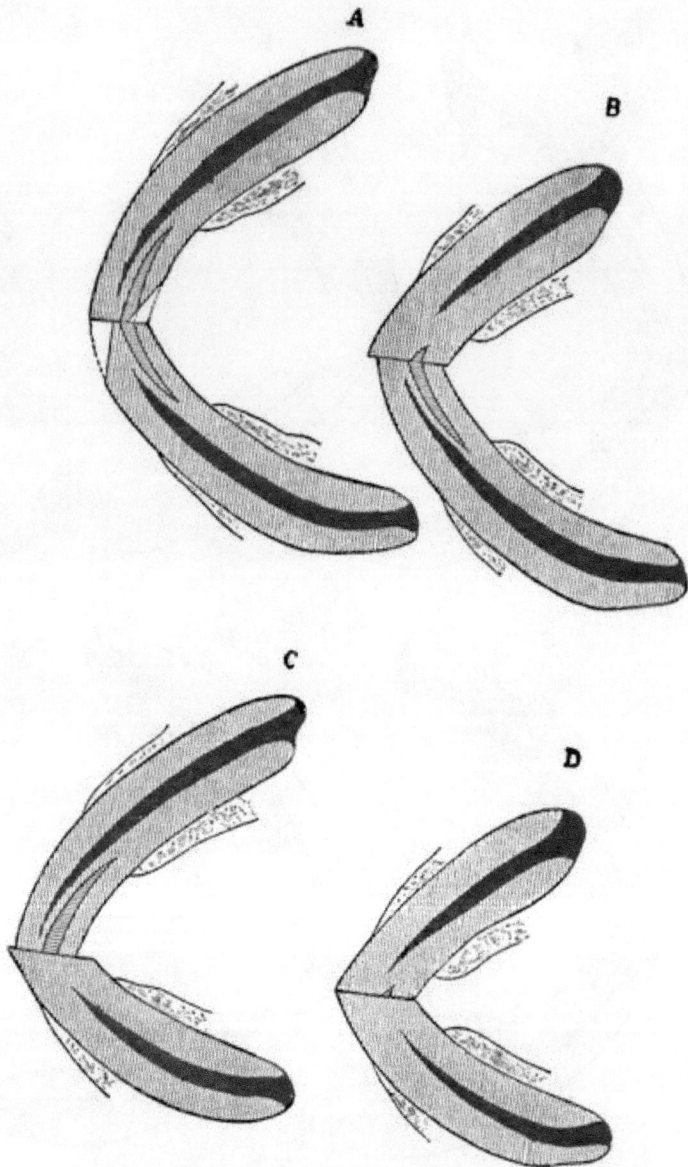

FIG. 155.

Crib-marks. A, the reverse of Fig. 154, D. B, C, and D, wearing of whole table,
upper, lower, or both.

thickness of the maxilla, the obliquity of the shortened teeth, and their size as they emerge from the gums (Fig. 155, B, C, D).

Fig. 156 shows the wearing produced by vertical objects, —halter-straps, which the animal renders tense by backing to its full length; wires, small chains, etc.

Most horses crib only on a given character of object, and when removed to another stall, or fastened so that they cannot reach the regular resting-place, may stop the habit for a time or be broken of it entirely. Others will, sooner or later, find some other body that suits them and commence again; so that the same animal may have two sorts of crib-markings. Others, when tied up, will learn to crib in the air.

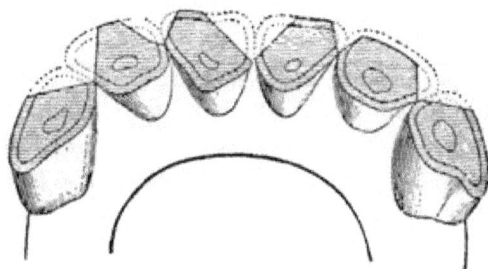

Fig. 156.
Cribbing on vertical edge of foreign body, or on halter-strap.

The rasp and file are sometimes used by dealers to shorten the incisors, and, while adding a little to the apparent age of the animal, remove the bevel which shows the habit of cribbing. A careful examination will always show a roughened surface, however, unlike the smooth polish of a naturally-worn table.

ARTIFICIAL IRREGULARITIES.

From time immemorial, the lower class of horse-dealer, like his congener in any other trade, has studied to learn and practice deceptive means for giving his articles for sale an apparent greater value than they really have. With this view, the dishonest breeder in the country attempts to hasten the appearance of age of his colts, to place them on the market as adult animals, and save a year's feed and cost of care when they are yet undeveloped; the city dealer fashions the mouth to destroy the evidences of cribbing, even at the risk of making the horse appear a year or two older, or alters the tables of the teeth to deceive the inexperienced buyer into thinking an old horse to be one just arrived at adult age.

Within recent years, moreover, especially in America, there has arisen a fraternity of "Equine Dentists," a guild endowed with great enthusiasm, who not only relieve the animals suffering from irregular and sharp molars, but, with artistic skill, remodel the whole mouth, and produce changes which sometimes greatly complicate the characters of the teeth as indications of the age of the horse.

REMOVAL OF TEMPORARY INCISORS IN ORDER TO AGE THE HORSE.

In Ireland, in Normandy (in France), in Virginia, and in some sections of the West, the temporary incisors are drawn some time before they would naturally drop to be replaced by the permanent ones, in order to hasten the eruption of the latter. If the intermediate incisors are drawn in a rising three-year-old, the permanent ones appear at three years, or soon after; and if the temporary corner incisors are then drawn, they are replaced in a few months by the permanent teeth; so that a rising four-year-old may have all of its permanent incisors, and the mouth of a four-year-old off may have the appearance of that of a horse a year older. The condition of the tush teeth does not control the age to any extent, as some horses have them between three and four years of age, and in others their eruption does not take place until six.

We have seen that when the incisors first appear they emerge from the gums somewhat obliquely, and later take their proper position, making their incisive arch a regularly rounded curve. When the permanent incisors have been hastened by the removal of the temporary ones, the former keep their oblique position, the arch is never regular, and is evidently diminished in width. If the removal has been recent the parts are inflamed, and there is sometimes a periostitis, which is evidently traumatic in origin. When such an irregularity is seen, the obliquity of the teeth in the curve of the incisive arch, a comparison of the two incisive arches (for frequently the deception has only been practiced on the lower teeth), and a comparison of the worn tables of the pinchers with the fresh edges of the others will be sufficient to show the fraudulent interference which has been executed. There are frequently cicatrices, showing the forcible removal of the teeth.

De Curnieu questions that the drawing of the temporary teeth hastens the eruption of the permanent ones; but Mayhew maintains the contrary, with which my own experience coincides. Mayhew says they can be hastened by the application of a hot iron to the gums. The question was put to a large number of breeders by MM. Goubaux and Barrier, and was answered, by all but one, in support of the opinion that the eruption of the permanent teeth is hastened by the removal of the milk-teeth.

BISHOPING.

Bishoping is a method employed by gyps to alter the appearance of the incisors, which can only deceive buyers who are entirely ignorant of the horse's mouth.

The crowns of the incisors of the young horse are wide from side to side; the dental tables are modified as the animal becomes older, and become successively oval, rounded, and triangular; the cups at first occupy the whole table, and are usually filled with dark-colored cement or black foreign matter; they gradually diminish in size, approach the posterior border of the teeth, and then disappear. In the centre of the table the dental star appears.

Bishoping consists in giving to the tables an artificial cup of a dark color. The teeth usually are first filed even; each table is then gouged out until somewhat concave, and the new cup is then blackened, either by nitrate of silver or by a point of white-hot iron. It is only practiced on the lower incisors.

Bishoping is readily recognized on a proper examination; as, with the shortened lower teeth, the tables of the two incisive arches usually do not correspond (Figs. 157 and 158), and the

FIG. 157.

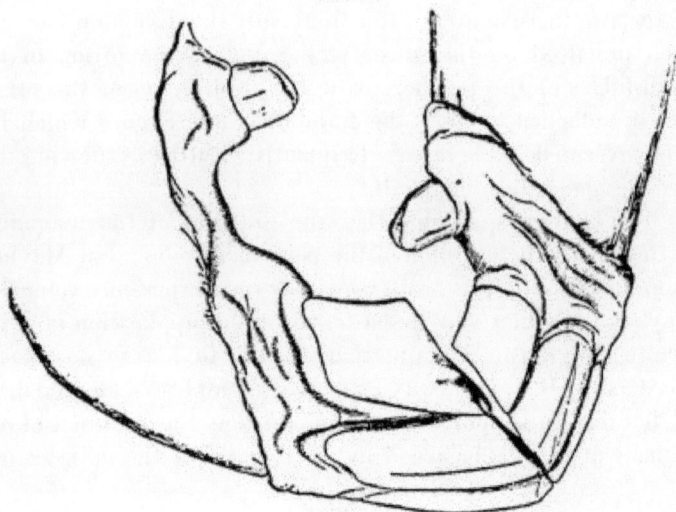

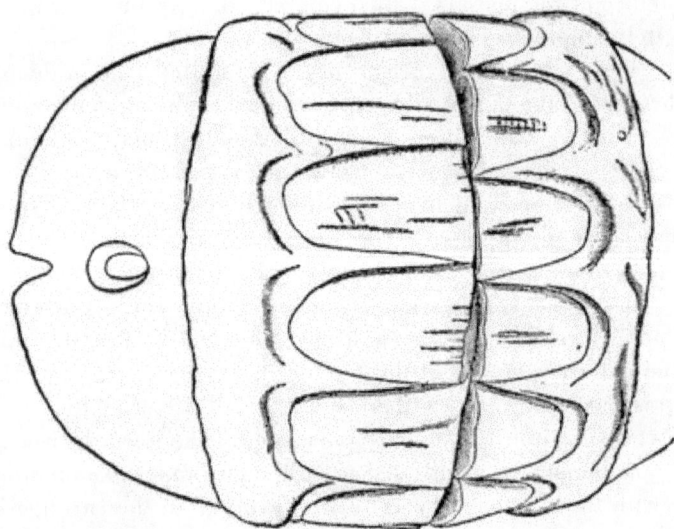

FIG. 158.

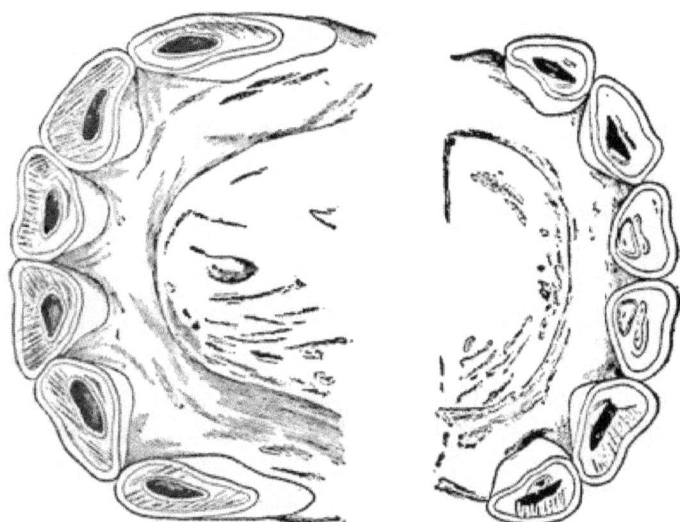

FIG. 159.

Bishoping (Figs. 157, 158, 159).—From in front the figure shows a space, above the corner and intermediate teeth, made by their having been filed. This is seen more plainly in profile. The tables show the roughened surface left by the file, and the artificial, blackened cups, at the posterior border of which are seen the real cups, surrounded by enamel.

enamel of the diminished cup of the horse is found posterior
to the artificial cup, or has disappeared (Fig. 159). In bishoped
mouths the artificial cup is found on the tables of wounded or
triangular teeth, in which they normally would not be present.
In bishoping the tushes are frequently filed down to point them
and make them appear fresh and small. This is evident from
the roughened surface and unnatural shape.

DRESSED MOUTHS.

The notch on the upper corner incisors and the whole
tables of the incisive arches are frequently filed down in "dress-
ing" the mouth. If recently done the roughened surface is
evident, and at other times it may complicate the appreciation
of age to a certain extent; but the incidence of the jaws, the
form of the tables, the conformation of the maxillary bones, etc.,
should prevent a deception of any importance.

AGE OF THE ASS, MULE, AND HINNY.

To determine the age of the ass, mule, and hinny requires an extension of the rules which have been applied to the horse. In the young animal the eruption of the milk-teeth and their replacement by the permanent teeth are about the same.

In these animals the incisors are smaller and narrower and less conical in shape than in the horse. The crown of the teeth is much longer and the root is shorter. The dental cup is continuous until a late period of life, as it is proportionately deeper. The cups are often imperfect behind. As the teeth are harder and more resisting, they wear much slower, and consequently the tables change their form more slowly. The teeth are more solidly imbedded in their alveolar cavities, and an excess of gum fixes them firmly. The dentine is more discolored and darker; the whole tooth is harder and resists the wear of dense, fibrous food better than that of the horse, and consequently does not wear so fast. The form of the tables is of less value, while the form of the arch and the obliquity of the teeth are more important. After seven years the elongated teeth and their approach to each other in a horizontal line, with thinning of the maxilla and the deposit of cement, the dark color and the narrowing of the incisive arches, are evidences of age.

The dental tables change from a round to a triangular form, which is complete at seventeen or eighteen years. At this age the cups may disappear and be replaced by the dental stars. The incisors become parallel to the maxillary bones and converge at their free extremities in very old age.

AGE OF THE OX.

THE age of the ox has demanded much less study than that of the horse, on account of its shorter life and the more limited time in which it is utilized for specific purposes. In a growing animal the evidences of its youth are unmistakable. Arrived at adult life, the difference of a year or two in its age is less important, in regard to its value, than it is in the horse, and later becomes even less so. The age of the ox is determined by the changes which take place in its dentition and the wearing away of its teeth, and by the changes in the growth, form, and appearance of its horns.

DENTITION OF THE OX.

Formula
$$\text{Temporary,} \quad \frac{0 \cdot 0 \cdot 3}{4 \cdot 0 \cdot 3} = 20$$

$$\text{Permanent.} \quad \frac{0 \cdot 0 \cdot 6}{4 \cdot 0 \cdot 6} = 32$$

The ox has thirty-two teeth,—twenty-four molars, arranged as in the horse, in arches of six on either side of each jaw, and eight incisors in the lower jaw, with none in the upper. In rare cases there are rudimentary molars (wolf-teeth), but when these exist in the young mouth they drop before the permanent dentition is complete. Tush teeth are not present.

INCISORS.

The upper jaw is devoid of teeth, but the intermaxillary bones are covered by a dense cartilaginous cushion and strong gum, which furnishes a resisting body to the incisors of the lower jaw in the prehension of food.

In the lower jaw there are eight incisors, arranged like the ribs of a fan, on the spatula-shaped arch formed by the extremity of the maxillary bone.

(158)

The incisors, instead of being fixed solidly in their alveolar cavities like those of the horse, are imbedded in them on a layer of cartilage, which allows of a considerable amount of motion, and thereby probably protects the cushion of the upper jaw from injury in seizing food, which is crushed rather than cut off.

The two middle incisors are known as *pinchers*, the next ones on either side as *first intermediate* teeth, the next as *second intermediate*, and the outside ones as *corner* teeth.

The two pinchers are slightly separated on account of the cartilaginous symphysis of the maxillary bones in the ox; this is much more marked in the first dentition. The other teeth touch each other by their extremities and form a complete arch, but, from their shovel shape, are not in contact along their borders as the wedge-shaped incisors of the horse are.

The incisors are composed of a crown and a root, separated by a distinct neck, giving them a somewhat shovel shape. The crown or free portion is flattened from above to below, and becomes (Fig. 160) thinner and broader at its anterior extremity.

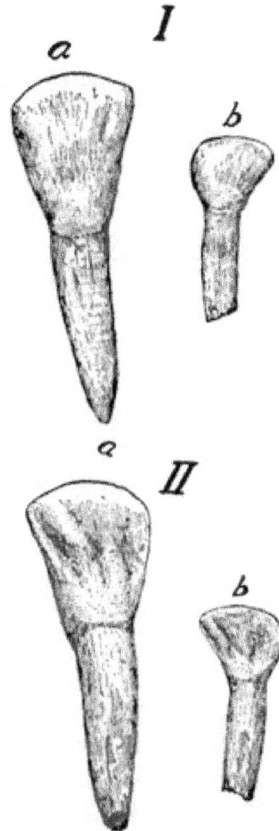

FIG. 160.
Left pincher teeth of the ox. I. External faces. II. Internal faces. *a a.* Permanent incisors. *b b.* Temporary incisors. (Natural size.)

The external or under face is convex in both directions; it is of a milky-white color and is striped with little longitudinal ridges and gutters, which become polished smooth with age. The internal or upper face is almost flat, but has a conical elevation, the base of which is directed toward the free border

with which it merges, while its sides are bordered by small
gutters.   The internal border is convex and the external border
is concave, which gives the tooth a curved shape and which indicates the side to which it belongs.   The root is rounded,
conical, and yellow in color; its extremity, in a young tooth,
shows the opening of the dental canal.

Structure.—The incisors have approximately the same
structure as the tush tooth of a horse; they are composed of a
dentine with the free portion covered by a continuous layer of
enamel.   The enamel is thickest on the external surface and
gradually disappears on the root.   In the young tooth there is
a large, simple, dental cavity filled with the dental pulp; but as
the animal gets older, a dark-yellow dentine is deposited until
the cavity is filled up and the tooth ceases to grow.   It is not
pushed from its alveolar cavity, as the incisors of the horse are,
as the free portion is worn away.

Like the horse, the ox has two sets of incisors,—the temporary or milk-teeth and the permanent ones.   The milk-teeth
are distinguished from the permanent ones from their being
smaller and narrower; their enamel is thinner and more trans-
parent and they are more curved to the side.   Their roots are
very short, and are pushed out by the replacing permanent
incisors.

The incisor scarcely reaches its full development when it
commences to be worn by its contact and constant friction
with food and with the cushion of the upper jaw.   The wearing
commences at the anterior border and removes the enamel
toward the posterior part of the upper face.   When it has
completely removed the conical eminence and the lateral gutters,
the tooth is said to be *leveled* and the table is formed.   From
the almost horizontal direction of the incisor teeth, the wearing
of their tables takes place in an oblique direction to their long
axis.

The table at first is large, consisting of a plate of dentine
surrounded by a border of enamel, and having in its centre a
transverse line of dark yellow, made by the uncovered, later de-

posit of dentine in the pulp-cavity. As the tooth becomes smaller toward its root the table becomes smaller and narrower and the dental star becomes narrower, but it also becomes longer from in front to behind, until it forms a distinct square as it is exposed by its posterior face from the oblique leveling.

As the incisors are worn away they seem to separate from each other at the roots, for the narrower parts of the crowns are held apart by their cartilaginous beds, which do not atrophy as does the bone of the maxilla of the horse, when the narrow, wedge-shaped roots are almost driven from their alveolar cavities.

When the teeth are worn down to their roots the gum retracts and shows the yellow stubs, which are all that is left of the incisive arch. (Fig. 167, b.)

With the wearing away of the incisors the dental arch loses its regular curve and becomes depressed in the middle.

MOLARS.

The ox has, like the horse, six molars in each arch, on either side of each jaw. The arch is shorter, as the teeth are smaller. The first molar is very small and each one increases in size to the sixth; but there is such a marked difference in the size that the first three teeth occupy but the third of the arch, while the last three complete the posterior two-thirds.

In the ox the molars have the same compound arrangement of enamel as in the horse, filled up by dentine and surrounded by a layer of cement. The latter often exists in great quantity, and is of a rich yellowish color. There seems to be a greater difference in the relative density of the substances, and the ridges of enamel stand out in sharper points as the softer dentine becomes worn away.

As in the horse, there are three temporary molars and six permanent molars in each arch.

According to Girard, the first temporary molar appears from the sixth to the twelfth day, following the eruption of the

11

second and third molars, which are sometimes through the gums at birth or appear immediately after birth.

The permanent molars make their appearance in the following order: Second molar from twelve to eighteen months; first molar from twenty-four to thirty months; third molar at thirty-six months; fourth molar at eighteen months; fifth molar at twenty-four to thirty months; and the sixth molar at three years or later. When the rudimentary molar (wolf-tooth) exists it appears about the tenth month, and is driven out of its alveolar cavity, with the appearance of the first permanent (fourth) molar, at the age of two years.

Simonds claims that the molars do not show in the calf at birth, and first make their appearance at the end of the first month. Simonds also differs slightly from Girard as to the eruption of the permanent molars, placing the appearance of the fourth molar at six months; the fifth at fifteen months; the sixth at two years. He evidently studied on very precocious cattle. Undoubtedly the eruption of the molars in cattle is variable, and, owing to the trouble of examination in the living animal, its careful study has been neglected.

An annexed table shows the diversity of eruption in various subjects. It will be found to vary somewhat, not only in the different races of cattle, but in the families of the same race which are reared under different climatic conditions and are nourished more or less liberally.

The periods of the animal's life, as indicated by the teeth, form the following natural divisions:—

1. Eruption of the temporary teeth.
2. Wearing of the temporary teeth.
3. Eruption of the permanent teeth.
4. Leveling of the permanent teeth.
5. Wearing away of the crowns.

TABLE OF ERUPTION OF THE TEETH IN THE OX.

	SIMONDS		GIRARD.	OTHER AUTHORITIES.		AVERAGE AUTHOR.	
	Race and Other Causes Favoring Development.	Race and Other Causes Retarding Development.		Minimum.	Maximum.	In Precocious Animals.	In Common Animals.
TEMPORARY INCISORS.							
Pinchers . . .			At birth.	Before birth.	3 days.	At birth.	At birth.
1st intermediate.			At birth.	Before birth.	5 days.	At birth.	At birth.
2nd intermediate.			5th to 9th day.	Before birth	12 days.	5 days.	12 days.
Corners . . .			13th to 19th day.	Before birth.	25 days.	12 days.	18 days.
MOLARS, 1st.	After birth.	1 month.	6th to 12th day.	At birth.	1 month.	6 to 12 days.	18 to 30 days.
2nd.	At birth.	Earlier.	Earlier.	At birth.	Few days.	At birth.	After birth.
3rd.	At birth.	Earlier.	Earlier.	At birth.	Few days.	At birth.	After birth.
PERMANENT INCISORS.							
Pinchers . .	1 y. 9 m.	2 ys. 3 m.	19 to 21 months.	15 mos.	2 ys. 3 m.	1 yr. 6 m.	20 mos.
1st intermediate.	2 ys. 3 m.	2 ys. 9 m.	2½ to 3 years.	2 ys. 3 m.	3 years.	2 ys. 3 m.	2 ys. 9 m.
2d intermediate.	2 ys. 9 m.	3 ys. 3 m.	3½ to 4 years.	2 ys. 9 m.	4 years.	3 years.	3 ys. 6 m.
Corners . . .	3 ys. 3 m.	4 ys. 9 m.	4½ to 5 years.	2 ys. 10 m.	5 years.	3 ys. 9 m.	4 ys. 6 m.
MOLARS, 1st.	2 ys. 6 m.		2½ to 3 years.	2 ys. 3 m.	3 years.	2 ys. 6 m.	
2d.	2 ys. 6 m.		12th to 18th month.	1 yr. 6 m.	2 ys. 6 m.	1 yr. 6 m.	
3d.	3 years.		3 to 4 years.	3 years.	4 years.	3 years.	
4th.	6 mos.		18 months.	6 mos.	1 y. 6 m.	1 yr. 6 m.	
5th.	1 y. 3 m.		2½ to 3 years.	1 yr. 3 m.	3 years.	2 years.	
6th.	2 years.		3 to 4 years.	2 years.	4 years.	2 ys. 6 m.	

DETERMINATION OF AGE BY THE TEETH.

With the young calf to be slaughtered for veal the absolute age is usually less important than its condition of development and its weight, but may be the cause of serious legal question and require the most acute perception on the part of the expert to decide and testify in controversies between the suspected butcher and the rigid law.

FIRST PERIOD—ERUPTION OF THE TEMPORARY TEETH.

The calf is sometimes born with no incisors; but usually the pinchers and first intermediate teeth have pierced through

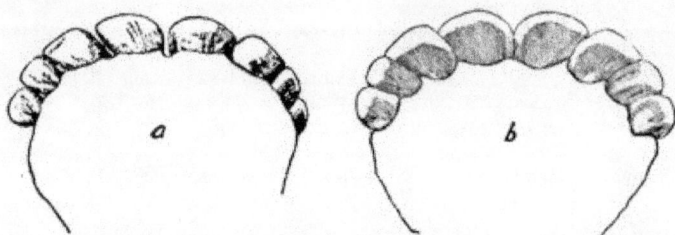

FIG. 161.
Incisors of the calf. a, external face. b, internal face.

the gums. The second intermediate teeth appear about the tenth day, and the corner teeth seven to ten days later, but may appear as late as the thirtieth day. The incisors do not reach the same level and complete the arch until the fifth or sixth month. (Fig. 161.)

SECOND PERIOD—WEARING OF THE TEMPORARY TEETH.

The leveling of the milk-teeth is very variable, according to the food on which the calf is fed. In calves fattened for the butcher with milk the wearing is slow, while in those that are put early to pasture and are fed on dried forage the leveling takes place much more rapidly. The pinchers are worn at their anterior borders at six months and are leveled at ten months; the arch is broken in the centre and loses its continuity at the first intermediate at one year, second intermediate at fifteen months, and at the corner teeth at eighteen to twenty months.

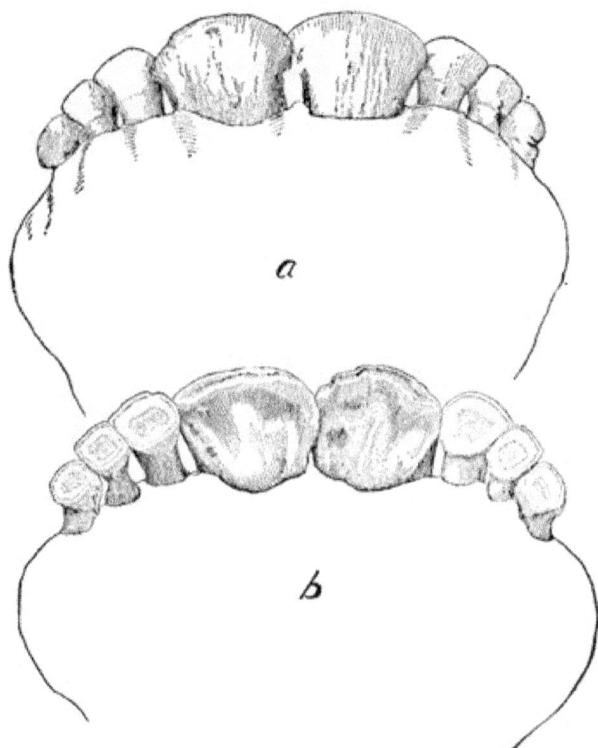

FIG. 162.

Two years. *a*, external face. *b*, internal face.

THIRD PERIOD—ERUPTION OF THE PERMANENT INCISORS.

Twenty Months.—At this time the milk-pinchers are replaced by the permanent ones, which protrude somewhat obliquely, but soon assume their natural position and are in place at two years. (Fig. 162.)

The crowns of the permanent pinchers, in this mouth, have not become quite free from the gums; the left-hand tooth is somewhat in advance of the other, and its enamel just shows traces of use.

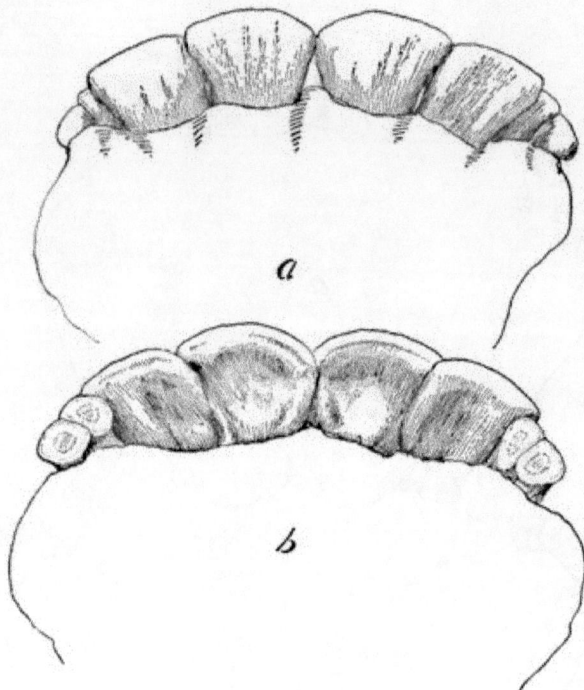

Fig. 163.
Two years, nine months. *a*, external face. *b*, internal face.

Two Years, Six Months.—Between two and one-quarter years to two years and nine months the first permanent intermediate teeth have accomplished their eruption. (Fig. 163.)

In this mouth the crowns of the first intermediate teeth are free from the gums, and the root of the second intermediate (temporary) is pressed on and its position slightly displaced.

Three Years, Three Months.—Between three years and three and one-half years the second intermediate have replaced the milk-teeth. (Fig. 164.)

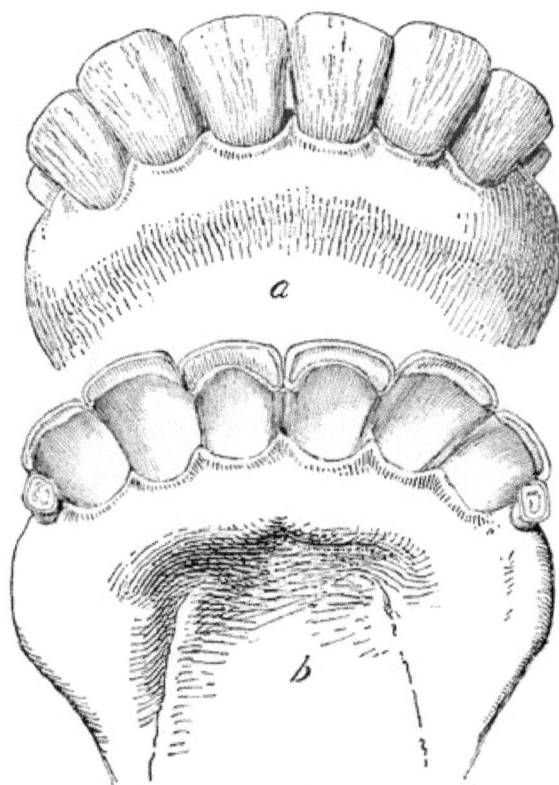

FIG. 164.

Three years, six months. *a*, external face. *b*, internal face.

In this mouth the second intermediate teeth have reached the level of the incisive arch and their enamel has commenced to be used.

Four Years.—Between three years and nine months and four years and six months the corner teeth are completely through the gums and the mouth is complete. (Fig. 165.)

The permanent corner teeth have just replaced the temporary ones, and are still in an oblique position, not having completely emerged from the gum.

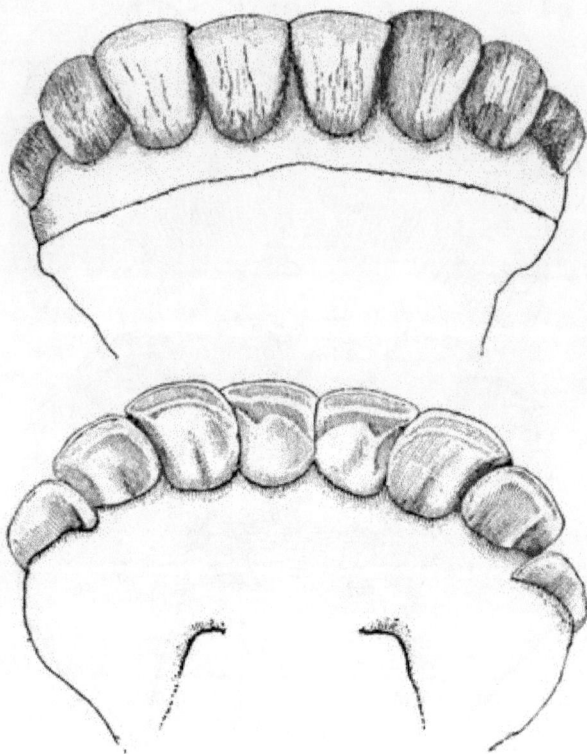

FIG. 166.
Four years. *a,* external face. *b,* internal face.

FOURTH PERIOD—LEVELING OF THE PERMANENT TEETH.

At five years the pinchers have commenced to level.

At six years the pinchers are leveled, both pairs of intermediate teeth are nearly so, and the corner teeth are somewhat worn.   (Fig. 166, *a.*)

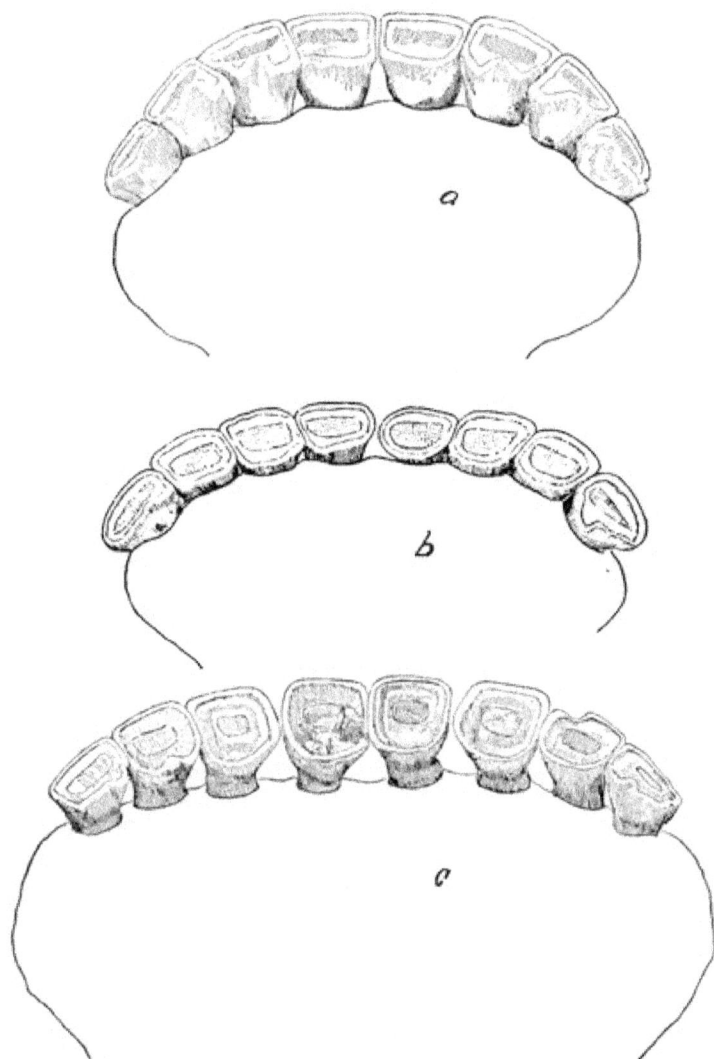

At seven years the first intermediate teeth are leveled, the second intermediate are much worn, and the corner teeth have lost their enamel at the anterior extremities.

At eight years the entire tables are leveled and the pinchers commence to show a concavity, which corresponds to a convexity of the cushion of the upper jaw.   (Fig. 166, *b.*)

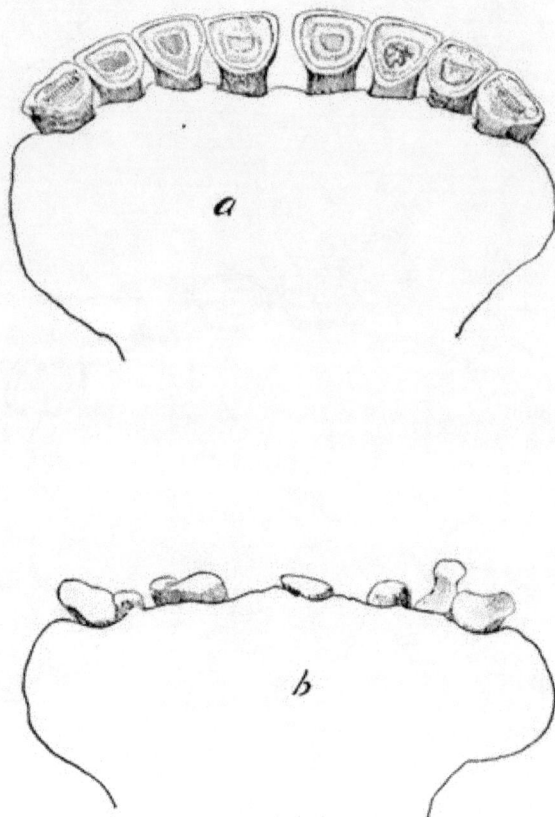

FIG. 167.
a, eleven years. *b*, fifteen years.

At nine and ten years this concavity extends to the intermediate teeth, the table of the pinchers is almost square, and the dental star of the pinchers and first intermediate teeth has become long and distinct.   (Fig. 166, *c.*)

During this period, from six to ten years, the rounded arch formed by the incisors gradually loses its convexity until it almost forms a straight line. The teeth appear to separate, and the gum shows between them.

FIFTH PERIOD—WEARING AWAY OF THE CROWNS.

From this time on there is a progressive change in the shape of the teeth; the crowns become worn down with more or less rapidity, they diminish in size, the dental stars become larger and square, the teeth seem to separate, and the retracting gum leaves the yellow roots uncovered.

At ten years the dental star is square in the pinchers, and the first intermediate and the corner teeth are leveled.

At eleven to twelve years the dental star is square in all of the teeth, which become triangular in shape and commence from this time on to be worn to stubs. (Fig. 167, a.)

It must always be borne in mind that the race of the animal and the character of the food produce great variations in the wearing of the teeth. Animals fed on hard forage and those fed on brewers' grains will have their teeth worn down much more rapidly than those fed on the prepared food of the ordinary dairy.

DETERMINATION OF AGE BY THE HORNS.

The horns of cattle, rising more or less gracefully from the frontal bones, were undoubtedly intended for weapons of offense and defense. All breeds of cattle are provided with horns, except that known as the angus or polled angus, which was indigenous to the northern part of Great Britain from the earliest historical times, but of which we have no trace in prehistoric deposits. Among all varieties of cattle individuals may be devoid of horns; they are known as "mulley," or, if deprived of their horns artificially (dehorned-dishorned), are called "polled."

Polling has been done by the Hindoos for over the last two thousand years, without ever showing trace of producing hereditary results.

The horns are symmetrical in shape, and when there is

any noticeable difference in the length, size, or curve of the horns, it may be assigned to some previous injury, disease, or accidental cause.

According to the character, habits, and surroundings of the various races and individuals, we find the horns more or less modified in size, shape, and strength.

In the semi-wild races, like those of Asia, Hungary, Spain, Texas, and South America, the horns may attain enormous size, —five feet in length; while in the most civilized, domesticated races, like the Durham, Dutch, and Channel-Island cattle, the horns tend to diminish in size, and may be but a few inches long.

The bull has strong, stout, short, straight horns, dense in structure, which seem to be as much points of hold for his massive, heavy head as actual weapons in times of warfare; the female has longer, sharper, more delicate horns, designed to use in emergencies. The steer has horns which are a compromise between those of the two sexes,—longer than those of the bull and larger than those of the cow.

The horns have for a basement two *cores*, or conical, bony projections of porous structure, richly supplied with blood-vessels, and containing air-cells which communicate with the sinuses of the frontal, occipital, and maxillary bones.

The cores are covered by a dense, fibrous, vascular membrane, from the outer face of which, corresponding to the chorion of the skin, the horns grow. The horns themselves are conical tubes more or less curved, consisting of concentric layers of epithelial growth.

Soon after birth the calf shows two little, hard, rounded points at either side of the frontal bone, which slowly emerge from the skin. At eight or ten days the point is through the skin and shows the color which the horns will have later; at three weeks a distinct little flexible horn has appeared. At five or six months the horn has commenced to curve on its long axis and assume the direction it will have later. Up to this time, and during the first year, the horn is covered by an epidermic prolongation of the skin, similar to the covering of the hoof of the foal at birth, but by the twelfth to fifteenth month this covering has dried and scaled off, leaving the natural, shining, tough surface of the horn proper.

In the second year the horns start a fresh growth, and a small groove is found encircling it between the substance secreted the first year and that which developed in the second.

During the third year a similar activity in growth takes place, and a second groove is found marking the line between the two years' growth. These two grooves or circular furrows around the horn are not well marked and have been frequently overlooked, and all trace of them disappears as the animal becomes older.

From three years on, the growth of the horn is marked by a groove or furrow, much deeper and so distinct that they show between them a decided elevation or " ring " of horny substance, which forms an accurate basis for estimating the age of the animal. In an animal over three years of age we count all of the horn beyond the first groove as indicating three years, and add one year to its age for each groove and " ring " which is present toward the base of the horn.

The grooves are always better marked in the concavity of the horn than on the convex surface. In feeble, ill-nourished animals they are but slightly marked.

Many causes, however, tend to diminish the value of the " rings " and grooves in the estimation of age. In " show " cattle and in herds of cattle kept for show, the horns are frequently sand-papered, scraped, and polished to give them the fine appearance of delicate texture, which, with that of the other integument, indicates the similar condition of the mammary gland for secreting milk and of the connective tissue for forming fat. Dealers scrape the horns to destroy the evidences of age in the animals which they have for sale. In old cows there is an atrophy of growth and an apparent contraction of the base of the horn; the rings and grooves are much less distinctly marked and may be indistinguishable.

In the first four years the teeth are the most valuable
indications of age, from four to ten years the horns furnish the
more accurate signs, and after ten years a careful comparison of
both is required to determine approximately the number of years
which have passed.

FIG. 165a.
Form of horns.   a, Hungarian; b, Jura; c, Jersey; d, Durham.

AGE OF THE SHEEP AND GOAT.

THE sheep is of value for any purpose for a still more limited period than the ox, and its exact age is of less importance, except in the case of valuable breeding animals, which are always possessed of a registered pedigree which guarantees the day of their birth. An approximation within a month or two is sufficient for practical purposes in a spring lamb, and an error of six months will not alter the taste of a four-year-old South-Down wether for a roast saddle. After this time it is a poor economist, except in districts where wool is the only product to be derived from the sheep, who will not turn it over to the butcher,—again excepting the pedigreed breeding animal. Civilization, better agriculture, and the care of man have altered the physiological characters of the ovine races even more than they have those of the bovine, and a larger percentage of the former have been rendered precocious than of any other species of animal. As a result of the improved agricultural needs, the sheep has altered considerably in form; the eruption of the teeth is more hasty, the horns have diminished in size or are absent, and the general shape of the animal has been modified.

The age of the sheep is determined by the character of the teeth, and of the horns, when the latter are present.

DENTITION.

Formula
$$\left\{ \begin{array}{l} \text{Temporary,.} \\ \\ \text{Permanent..} \end{array} \right.$$
$$\frac{0 \cdot 0 \cdot 3}{4 \cdot 0 \cdot 3} = 20$$
$$\frac{0 \cdot 0 \cdot 6}{4 \cdot 0 \cdot 6} = 32$$

The sheep has thirty-two teeth, like the ox,—eight incisors in the lower jaw, none in the upper, six molars in each arch of either jaw, making twenty-four molars in all, and no tushes.

INCISORS.

The incisors are eight in number. They are set firmly in their alveolar cavities in the maxilla, and form an arch more convex than either the incisive arch of the horse or the ox. As in the ox, they are termed the *pincher, first intermediate, second intermediate*, and *corner* teeth. As in the other animals, there are two sets, the *temporary* incisors and the *permanent* incisors. The temporary incisors are much smaller and proportionately much narrower than the permanent ones; so that the jaw of the lamb has an elongated, narrow appearance, which alters greatly in form, becoming wider and more flat in the older animals. The incisors of both dentitions are wedge-shaped like the permanent incisors of the horse. They have no neck separating the crown from the root, like the incisors of the ox and the temporary incisors of the horse. They are firmly imbedded in their alveolar cavities, which allows them to nip the short grass close to the roots and obtain a living, where the ox, with its loose incisors, can no longer obtain a hold to tear up the blades.

In the virgin tooth the external face is white and polished except near the root, where it is surrounded by a black cement. The internal face has two longitudinal gutters divided by a little crest. The gutters are filled with black cement. They represent by their convex anterior face, in profile, the quarter of a circle. From this position they meet the cushion of the upper jaw by their free extremity, like the incisors of a young horse, and not by the posterior surface, as in the ox; so that by use they rapidly wear the anterior border and form a table to the teeth like the soliped, and not like that of their closer relation, the large ruminant. The incisors of the sheep are formed of dentine, surrounding a pulp-cavity which becomes filled with a darker-colored deposit at an early period, and are covered with a layer of enamel, which disappears toward the roots, and which is covered on the sides, in the longitudinal gutters, and near the gums by a black cement.

MOLARS.

The molars, except for their smaller size, resemble those of the ox. They gradually increase in size from the first to the sixth molar, the first three occupying but a third of the arch. The relative density of the dentine and the convoluted enamel differs greatly, so that the dentine is more rapidly worn and the enamel stands in sharp ridges from the table of the teeth.

ERUPTION OF THE TEMPORARY TEETH.

INCISORS.

Considerable diversity of description is found among the older writers as to the time of eruption of the teeth in the sheep from the fact that some studied them in the common races of Central Europe, while others examined them only in the perfected races of England. At the present day the studies of the latter will be found more accurate, as few flocks can be found which are not well bred or mixed more or less with the finer races.

At birth, the lamb may have the pinchers and first intermediate teeth through the gums, or the anterior borders of these teeth may show a whitish line where they press against the gum, which they pierce in from three to five days. The second intermediate emerges from the gum about the tenth day, and is followed tardily by the corner teeth, which do not appear until the twenty-eighth or thirtieth day.

MOLARS.

The three temporary molars in each arch appear about the third week, when we find the first six incisors sufficiently advanced from the gums to allow the young animal to nip herbage, which is to be ground by the larger teeth.

ALTERATIONS DURING THE FIRST YEAR.

One to Three Months.—During this time there is little change in the teeth. They are slightly more prominent, but retain their virgin appearance.

12

About Three Months.—Sometimes just before and sometimes just after three months from the birth of the animal the fourth (first permanent) molar appears, that of the lower jaw preceding that of the upper by a week or more.

Three to Nine Months.—During this half-year no characters of substantial value can be found. The incisors during this period attain their full size, and, according to the herbage or other food on which the animal has been fed, more or less wearing of the anterior borders of the teeth has taken place. In addition, however, to the general aspect of the animal, it is easy to recognize the difference between the fresher teeth, which have scarcely been worn, of a lamb four to five months of age, and the worn, broken, and loosening ones at eight, nine, or ten months.

Nine Months.—At nine months the fifth (second permanent) molar makes its eruption. This fixes an important land-mark in the determination of the age of the young animal.

The fourth and fifth molars are formed of two principal lobes, each with two points of enamel, so that there are four points of enamel; the internal points are higher on the lower teeth, while the external ones are longer on the upper teeth. These points become used, and as they are worn away the band of central enamel is separated from the outer border of the teeth by a larger area of dentine.

The character of the points of enamel on these teeth, or the distance of the bands of enamel from the outer border of the teeth, thus indicates whether the animal has just passed nine months or is a year or more old.

Ten Months.—At about ten months the body of the maxilla is seen to be wider and broader,—due to the development of the permanent pinchers, which press upon the roots of the temporary pinchers, producing absorption, and these latter are found less solidly fixed in their alveolæ.

ERUPTION OF PERMANENT TEETH.

One Year to Fifteen Months (Fig. 168, I).—At this period the permanent pinchers make their appearance. In the more

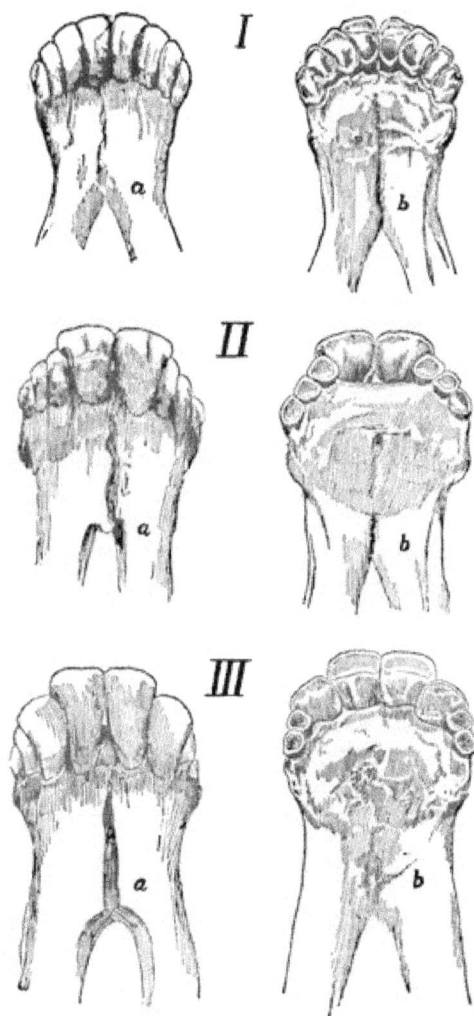

FIG. 168.
I, one year. II, eighteen months. III, two years, three months. *a*, external face. *b*, internal face.

precocious races the eruption of the pinchers takes place at one year, while in the common races it may not until fifteen months. Again, a question of individual development may hasten or retard the eruption by a month or two; so that the better-bred animal may be thirteen or fourteen months before the first two permanent teeth have appeared, while the more hasty, common animal may have its permanent teeth by the same time. In the male the eruption takes place very slightly earlier than in the female.

The conditions of the points on the fourth and fifth molars already alluded to will aid in deciding if the eruption has been hasty or tardy.

Eighteen Months (Fig. 168, II).—At a year and a half the first intermediate permanent incisors and the sixth permanent molars appear. The eruption of these teeth is followed very shortly by the falling out of all the temporary molars, which are replaced almost simultaneously by the permanent molars. After two years the molars furnish but little indication of the age. From this time, according to its sex, the lamb takes the name of *ram* or *ewe*.

Two Years and Three Months (Fig. 168, III).—About nine months after the eruption of the first intermediate teeth the second *intermediate* appear ; whether in precocious or tardy races, the interval between the eruption of the first and second intermediate teeth is some three months longer than that between the former and the pincher teeth. In tardy animals, in which the first intermediate have not appeared until they are nearly two years of age, the second intermediate may be delayed until two years and nine months.

Three Years (Fig. 169, IV).—At three years the eruption of the corner teeth takes place, although under the conditions given above certain animals may be several months late in completing their dentition.

A greater irregularity is found in regard to the appearance of the corner teeth than in the sequence of eruption of the others. By the time of the appearance of the corner teeth the pinchers and first intermediate may become worn and even leveled. Girard first noted that some sheep have but six incisors, the corner teeth remaining aborted under the gum.

After four years (Fig. 169, V) there is a continuous but very irregular wearing of the teeth, which is governed by the nature of the food. Those animals which are fed on fine herbage and those which are stall-fed level off the tables evenly until but stumps remain ; but in all sheep there is always more rapid wearing of the pinchers and first intermediate teeth, which gives the plane of the tables of the arch a concave form.

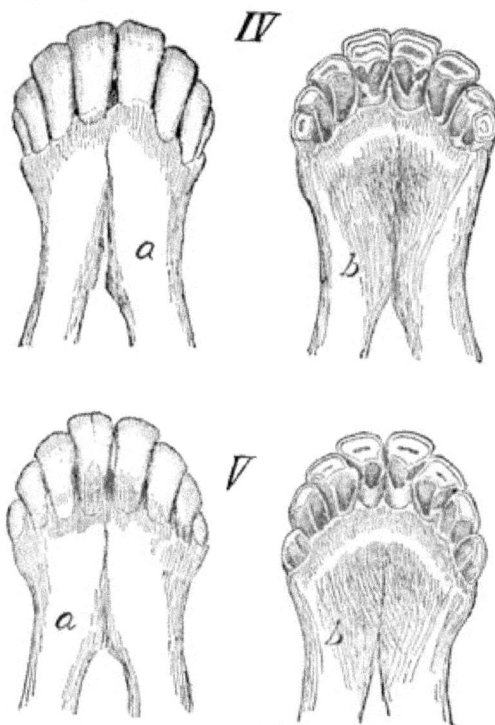

FIG. 169.
IV, three years. V, four years. *a*, external face. *b*, internal face.

Those animals fed on the hill-sides and where shrubs and undergrowth are abundant use their teeth much more rapidly, and wear notches between them like the markings on a horse which cribs on a vertical object ; these notches are called, very appropriately, by the French agriculturists, *swallow-tails* (*queue d'hirondelle*).

As age advances the teeth of the sheep are worn to stumps; they become black and loose in their alveolar cavities.

In the old animals the face wrinkles, the lips become thick, and the muzzle, which was fine and pointed in the lamb, becomes enlarged and broad. Sometimes in old sheep, but more frequently in old goats, the incisors attain a great length; the incisive arch is, however, in these cases usually broken and irregular.

DETERMINATION OF AGE BY THE HORNS.

The horns of the ovine and caprine races are variable ornaments. Some races have horns, some do not; in some races the male has horns and the female none, while in the others the female has very small, aborted apologies for them. In the long-wooled English sheep the horns are absent; the merinos originally had the magnificent typical "ram's horn," which has become a descriptive of shape, but domestication has gradually reduced them in size, and in some families they abort entirely.

The horn of the sheep grows up, out, back, and gradually turns on a central axis, according to its length, in the form of a conical corkscrew. The horn of the goat grows up, back, and slightly outward.

In structure and growth the horns of the sheep and goat are like those of the ox. The horn of the sheep is flat on its inferior surface and convex from side to side on its superior surface; it is divided transversely by ridges; the interspaces have longitudinal scaly ridges and gutters.

The horns appear about fifteen days after birth, attain their greater growth during the first year, and cease growing after four years. If the animal is castrated the growth of the horn lessens, whereas in the ox castration stimulates the growth.

The epidermic covering, which extends from the skin on the appearance of the horn, dries and scales off in six weeks to two months, leaving a roughened, scaly surface. The size and

roughness of these series of scaly sections vary from year to year,
and indicate the years until the horns cease to grow,—at four.
Observations by Girard on the growth of the horns of merino
rams (the horns of the merinos are the most typical) showed the
growth to be :—

The first year,	. . . .	19 to 20 inches.
The second year,	. . . .	5 to 6 "
The third year,	. . . .	3 to 4 "
The fourth year,	. . . .	2 to 3 "
		29 33 "

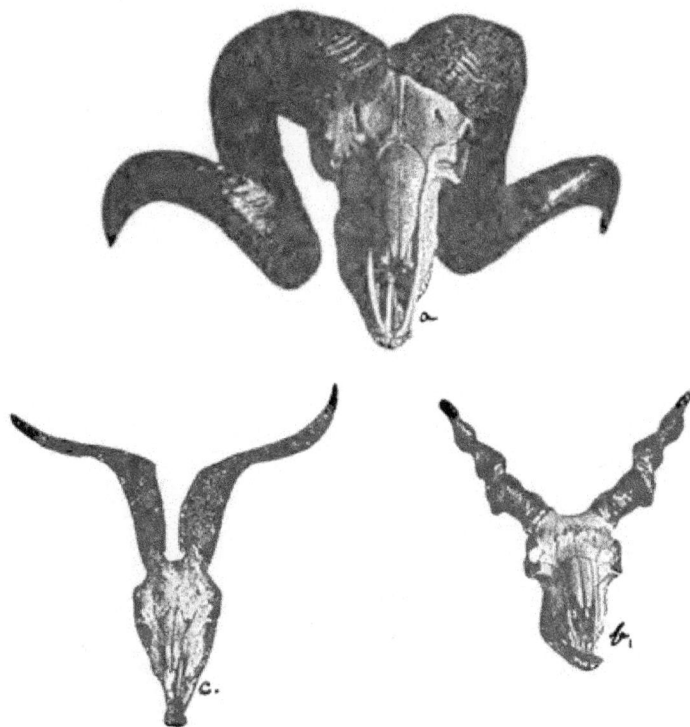

FIG. 109a.
a, Merino ram ; b, Montenegrin ram ; c, goat.

AGE OF THE HOG.

THE age of the hog is of much less importance than that of any of the domestic animals, as this animal becomes an article of commerce and is sent to the butcher the moment that the ratio of its increase of weight (flesh and fat) ceases to be greater than the proportionate cost of the food which is furnished to it. Even for breeding purposes, the hog is rarely kept to an advanced age. The question of its age only becomes important in exhibitions, where the judges must verify the age in the class for which the animal is entered.

DENTITION.

Formula $\begin{cases} \text{Temporary,} \qquad\qquad \dfrac{3 \cdot 1 \cdot 3}{3 \cdot 1 \cdot 3} = 28 \\[2ex] \text{Permanent,} \qquad\quad \dfrac{3 \cdot 1 \cdot 4 \cdot 3}{3 \cdot 1 \cdot 4 \cdot 3} = 44 \end{cases}$

The hog has forty-four teeth,—six incisors in each jaw (12), two tushes in each jaw (4), and seven molars (which are subdivided into four premolars and three molars) in each arch of both jaws (28).

INCISORS.

The pinchers and intermediate teeth of the upper jaw are flattened from side to side, somewhat like those of the horse, as they also have a small dental cup; while those of the lower jaw are straight, rounded, and directed forward like the teeth of a rodent, with their extremities converging toward the median line. The corner teeth of both jaws are small, conical teeth, separated from the intermediate on the one side and from the tusks on the other. They have no cups, but often have three small tubercles like the incisors of the dog.

(184)

TUSKS.

The tusks are much more developed in the male than in the female; they are prismatic in shape and curve outward and backward. In the wild hog they may attain an enormous length (six inches or more), pressing the upper lip upward and the lower lip downward and crossing each other in an **X**, the superior tusk passing behind the inferior; the two frequently have their respective anterior and posterior faces much worn by the constant friction from their contact with each other. The tusks continue to grow during the life of the animal. The growth is diminished in the castrated male.

MOLARS.

The molars are divided into premolars and post-molars. The first premolar resembles very much the conical corner incisor tooth, but has two roots instead of one. It is placed in the interspace between the tusks and the other molars, being about twice as far from the latter as from the former.

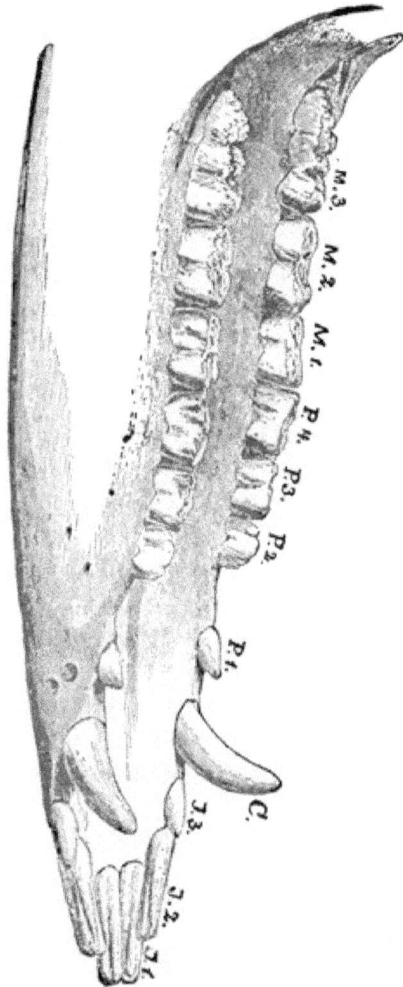

Fig. 170.
Lower jaw of pig, showing permanent dentition.

The other molars are between the inferior molars of the horse and the molars of the dog in form. They wear by the centre of the crown and have not the irregular asperities found in the herbivora. They increase gradually in size from in front to behind, and are larger in the upper jaw than in the lower.

DETERMINATION OF AGE BY THE TEETH.

To Dr. Olof Schwartzkopff I am indebted for the following:
"During the past few years many objections have been raised, on the part of our practical breeders, to the correctness of the older rules for recognizing the age of our domestic animals. Several cases of an extraordinarily early development of the dentition have been observed in fat-stock shows and other exhibitions, and it has been alleged that modern feeding, with the tendency to produce early maturity, results also in an earlier shedding of the teeth. Not only in the United States have these doubts been heard, but also in England and Germany. In 1882 Prof. G. T. Brown published, in the *Journal of the Royal Agricultural Society of England*, an article in which he comes to the conclusion that, as a general thing, the views of the breeders cannot be relied upon, and that the recognition of the age from the teeth is still the best and surest. In June, 1886, the Executive Committee of the Fat-Stock Show at Berlin preferred similar complaints, and requested the Minister of Agriculture to introduce new experiments at the veterinary schools and agricultural experiment-stations in Germany, to ascertain whether the signs of age from dentition, sexual development, and growth of horns can appear at an earlier time in our precocious breeds than hitherto believed. Accordingly, Prof. A. Nehring, of Berlin, published, in the *Landwirtschaftliche Jahrbücher* of 1888, a series of new dentition tables for pigs, as a result of his studies and investigations upon a collection of one hundred and thirty-one skulls of different kinds of pigs, at the Museum of the Royal Agricultural School at Berlin.

"Having seen and examined parts of this collection, I will

undertake to demonstrate, with the guide of the tables mentioned, together with my own experience and observation while practicing in breeding establishments, that our practical men have been quite right in many cases, and that the doubts to which reference has been made are not without foundation."

The pig, like the other animals, has two sets of teeth,—the temporary or milk teeth and the permanent ones.

"There exists a remarkable difference in the time occupied by the teeth in cutting their way through the gum and appearing on its surface, while the mode of succession remains unchanged. But it must be remembered that the dentition tables, still referred to in modern books for the practical pig-breeder, are based upon observations made in times when the common pig was raised, or, perhaps, a breed more or less improved by English stock, and fed in the old-fashioned style. Variations into early maturity were then described as abnormal; but as soon as the pure breeding of the favorites of our day commenced, Berkshire, Poland China, *et al.*, and we applied to them scientific feeding, we forced the animals into entirely new and artificial conditions, revealing the hitherto unknown physiological laws of early maturity."

The periods of age which can be determined by the dentition are divided as follows:—

1. Eruption of the temporary teeth.

2. Eruption of the permanent teeth.

3. Wearing of the permanent teeth, after two and a half or three years,—an age attained but by few animals.

These periods vary slightly, and when a latitude of time is given it is understood that the shorter is for precocious pigs and the longer for those races which approach the primitive pig.

FIRST PERIOD—ERUPTION OF THE TEMPORARY TEETH.

At Birth.—The pig is born with eight teeth,—the corner incisors of each jaw and the tusks. These teeth resemble each other very much, and probably serve to aid the tongue in sucking.

Four to Fourteen Days.—In the first two weeks the second upper molar and the third lower molar appear.

Two to Five Weeks.—During the next three weeks the second lower molar, the third upper molar, and the pinchers of both jaws accomplish their eruption.

Five Weeks to Three Months.—At about six weeks the intermediate incisors of the lower jaw appear, which are followed in two weeks by the intermediate incisors of the upper jaw.

Three Months.—At three months the eruption of the temporary teeth is completed, and during the next few months the teeth become more or less worn, according to the character of food upon which the animal is nourished.

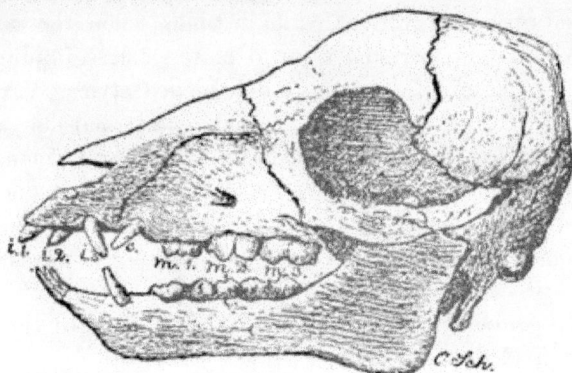

FIG. 171.
Skull of a three-month-old pig with full milk dentition.

SECOND PERIOD—ERUPTION OF THE PERMANENT TEETH.

About Six Months.—Before the first half-year the first permanent molar (fifth) and the first premolar (wolf-tooth) in each jaw appear.

Nine Months.—Within a month of three-quarters of a year the corner incisors and the tusks make their eruption, followed shortly by the second molars (sixth) of both jaws.

One Year.—At one year the lower pinchers protrude from the gums, preceding the superior pinchers by two to three months. At about this time, thirteen to fourteen months, the third and fourth premolars appear, followed a month later by the second premolar.

One and a Half Years.—At eighteen months the intermediate incisors of both jaws make their appearance, and simultaneously the eruption of the third molar (seventh) takes place.

After Two Years.—After two years the incisor teeth become used with great irregularity, according to the nature of the food; the tusks increase in length and give some approximation of their age, but insufficiently to be accurate.

TABLE OF DENTITION OF THE HOG.

MILK DENTITION.

| TEETH. | SCHWARTZKOPFF | | PRIMITIVE PIGS. | SIMONDS. |
	PRECOCIOUS PIGS.	NORMAL TIME OF APPEARANCE.		NORMAL TIME.
Corners and tusks.	Present at birth	Present at birth.		At birth.
Pinchers	2 weeks.	3 to 4 weeks.	5 weeks.	1 month.
Intermediate :				
Upper jaw . . .	8 weeks.	12 weeks.	16 weeks.	} 3 months
Lower jaw . . .	5 weeks.	8 weeks.	12 weeks.	
1st molar . . .	5 weeks.	7 weeks.	9 weeks.	
2d molar :				
Upper jaw . . .	4 days.	8 days.	14 days.	
Lower jaw . . .	2 weeks.	3 to 4 weeks.	5 weeks.	
3d Molar :				
Upper jaw . . .	2 weeks.	3 to 4 weeks.	5 weeks.	
Lower jaw . . .	4 days.	8 days.	14 days.	

PERMANENT DENTITION.

Pinchers	11 months.	12 months.	14 months.	12 months.
Intermediate .	16 months	18 months.	21 months.	18 months.
Corners	7½ to 8 months.	9 months.	10 months.	9 months.
Tusks	3½ months.	9 months.	10 months.	9 months.
1st premolar . . .	2 to 3 months.	5 months.	6 months.	
2d premolar . . .	13 months.	14 to 15 months.	16 months.	
3d premolar . . .	12 months.	13 to 14 months.	15 months.	
4th premolar . . .	12 months.	13 to 14 months.	15 months.	
1st molar	2 months.	5 months	6 months.	
2d molar	7 to 8 months.	9 to 10 months.	12 to 14 months.	
3d molar	17 months.	18 to 19 months.	20 to 22 months.	

The following explanation of the causes influencing the variations in the dentition of the pig is given by Schwartzkopff:—

"The question now arises as to what may be regarded as the cause of this early dentition in modern pigs. We know that our present method of feeding causes a rapid development of the whole body, including, of course, the head. As the teeth could not possibly grow without a corresponding growth of the jaws that produce them, we must conclude that the development of the skull is the primary cause or driving force in their development. Unconsciously, the modern feeder has produced here some highly interesting facts instructive to natural science at large.

"Hitherto zoölogists have been of the opinion that the form of skull of a fixed species is unchangeable from generation to generation,—we may say for thousands of years. This is correct as long as we think of individuals raised in the freedom of nature and under natural and similar circumstances. But domestication, with its forced feeding and breeding for various demands, has brought about unexpected changes in many respects, and it is now evident that the form of skull does not rest merely upon heredity. Only a predisposition to a certain form of skull is transferable from parents to their offsprings, but whether exactly the same form will be transmitted depends to a greater extent upon the nutrition, and but little less upon the employment of the muscles of the head and neck. It is not only important that the nourishment be abundant and well selected, but it is also necessary that the individual be in a healthy condition, and his digestive apparatus in such working-order as to be able to utilize the offered food equally well. This is plainly seen by comparing skulls from animals which were healthy and growing vigorously with those which received the same advantages of nutrition, but were suffering with a chronic disease. Continued weakness, caused either by disease or insufficient food, produces a long, slender skull, while the skull from a strong pig shows a remarkable expansion in its latitude and altitude.

"The following reproductions, taken from originals in the agricultural museum at Berlin, will illustrate this point:—

" Besides the nutrition influence, a strong or weak muscular action plays an important part in the production of form. The pulling and pressure of muscles extensively used for certain purposes, especially those of the head and neck, will give the head a characteristic shape. Pigs which are prevented from rooting will acquire a short, high, and rounded head, while those which are forced to root to secure a portion of their food will develop a long and slender form of head. If we force both experiments to the greatest degree possible, we shall produce

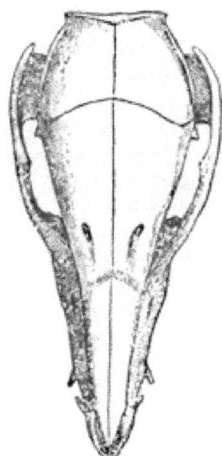

FIG. 172.
Skull of a three-month-old pig which died from tuberculosis. (Half natural size.)

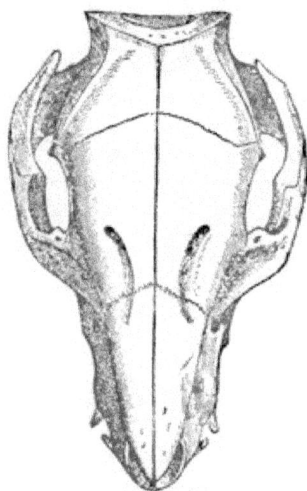

FIG. 173.
Skull of a two-month-old healthy and well-fed pig. (Half natural size.)

those extremes which distinguish the wild pig from our improved races. That this is true is proven by the fact that when our domestic hogs are returned to absolute liberty, it will require but a few generations to reproduce the original skull of the wild pig. And, *vice versâ*, we have called into existence, from the primitive hog, all those different representative types of our day by careful and continued selection, gradual assortment, and particular attention to the desired qualities of form, size, etc. The striking difference between the skull of a primitive hog and a modern one is seen in the following illustration :—

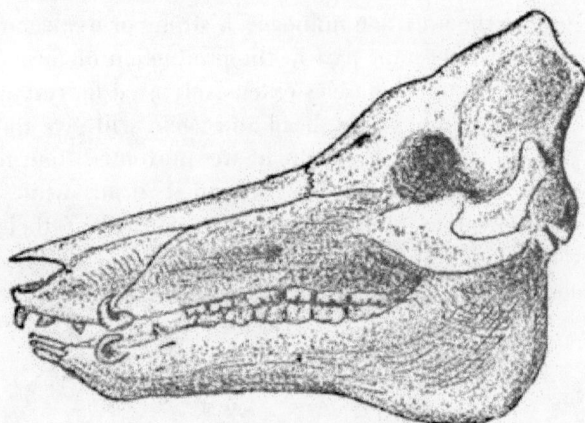

FIG. 174.
Skull of a full-grown primitive Texas boar.

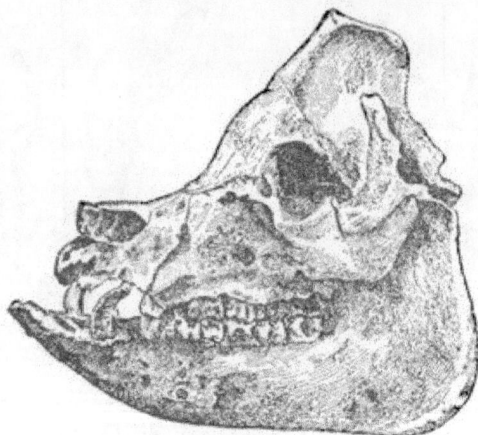

FIG. 175.
Skull of a full-grown sow of the small Yorkshire breed.

"The pig has, perhaps, the most elastic and changeable organization of any of our domestic animals. It also has the advantage of being able to digest all kinds of food as an omnivorous animal; and last, though not least, it multiplies more rapidly than any domestic animal,—even the sheep. Therefore, it has been at all times regarded, and properly too, as the animal *par excellence* for experiments in breeding, and the pig is the best example of what men have accomplished in the production of animals.

"Drawing, now, the conclusions from the above examinations, I shall summarize them in the following theses :—

"1. The order of succession of the teeth in our precocious pigs remains the same as in the primitive hog.

"2. The times when the teeth appear are variable, according to the race, feeding, and health. The same breeds, raised under the same conditions, will show the same appearance.

"3. The form of the skull depends upon nutrition, health, and more or less employment of certain muscles of the head and neck."

13

THE age of the dog is a matter of considerable importance. While the well-bred hunting-dogs, hounds, setters, and pointers, and those of other races which are kept for breeding purposes, have almost invariably a verified and registered pedigree, which is stereotyped on the end of the owner's tongue, with the date of the animal's birth, yet there are large numbers of dogs, especially of the races which are kept for house-dogs and pets, which, from their appearance, are of undoubted good breeding, but which, from change of owners by sale, theft, or in the many other ways peculiar to dogs, lose their records. In view of the great prices which these animals sometimes bring, it becomes needful to at least approximately determine their age, and to estimate the number of years which they still have to live and will be of use to their owners. The smaller races of dogs usually live to a greater age than the larger. The mastiff, St. Bernard, and Great Dane rarely exceed the age of ten years; the hound, setter, and pointer may live two or three years longer; and the terriers have been known to live to seventeen, nineteen, and twenty-two years of age.

DENTITION.

$$\text{Formula} \begin{cases} \text{Temporary,} & \dfrac{3 \, . \, 1 \, . \, 3}{3 \, . \, 1 \, . \, 4} = 30 \\[2ex] \text{Permanent,} & \dfrac{3 \, . \, 1 \, . \, 4 \, . \, 2}{3 \, . \, 1 \, . \, 4 \, . \, 3} = 42 \end{cases}$$

The dog has forty-two teeth,—twenty in the upper jaw and twenty-two in the lower. They are divided, like those of the other animals, into *incisors*, *tusks*, and *molars*, the latter again subdivided into *premolars* and *post-molars*.

(194)

INCISORS.

There are six incisors in each jaw, known as the *pinchers*. *intermediate*, and *corner* teeth. Those of the upper jaw are larger than those of the lower; the corner teeth are the largest, the intermediate are next in size, and the pinchers are the smallest. The incisor teeth have a large crown, on the extremity of which there are three eminences or tubercles, that in the centre being the largest, which gives it somwhat the form of a clover-leaf, or *fleur-de-lis*. The internal face is beveled off and separated from the root of the tooth by a distinct ridge. The external face is convex in both directions, and is smooth and shiny. The root is very large, it is flattened from side to side, and is separated from the crown by a well-developed neck. The root is very firmly imbedded in the deep alveolar cavities. The virgin root contains a large dental pulp-cavity, which, however, becomes obliterated at an early period. When the incisors are worn by the friction of food and the thousand-and-one foreign bodies which the dog "handles" with its mouth, the central tubercle is first used, then the lateral tubercles, and last the whole crown is worn, showing an irregular table of dentine, surrounded by enamel, and having in its centre a dark spot corresponding to the obliterated dental cavity.

The surface of the teeth is of a brilliant white, as, in the dog, they are not covered with cement as in the other animals. The temporary and permanent incisors are alike in form and shape, but the former are very much smaller than the latter, and, as the head and jaw of the young dog are relatively large, the temporary incisors cannot occupy the whole arch, and so have spaces between them showing the gum, while the larger, permanent teeth, when virgin, are just in contact with each other by their free extremities.

TUSKS.

The dog has four tusks, which are very prominent and give the name of *canine teeth* to the corresponding teeth which resemble them in the other animals. They are placed one each side of each jaw, dividing the space between the incisive arch

and the arches of the molars, unevenly ; the inferior tusks are
nearer to the incisive arch, while the superior ones are nearer
the molars, and when the jaws are closed the tusks cross each
other, coming in contact by the postero-external surface of the
inferior tusk and the antero-internal face of the superior tusk.
The temporary tusks are smaller, longer, more curved, and more
pointed than the permanent ones.

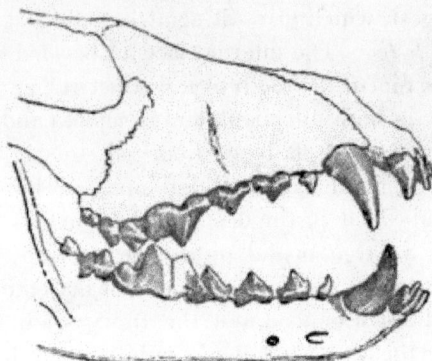

FIG. 176.
Permanent dentition of the dog.

MOLARS.

There are six molars in each arch of the upper jaw and
seven in each arch of the lower. They increase in size from
the first premolar to the fourth of the upper jaw, and to the
first post-molar (fifth of the lower jaw), and then diminish in
size at the next to last and last molar.   They have large,
bulbous crowns with pointed eminences, which serve for the
tearing rather than the grinding of food.   In the upper jaw
the first three premolars are unilobular, the last is bilobular,
and the last two post-molars have flat crowns.   In the lower
jaw the four premolars are unilobular, the first molar has three
eminences, and the last two have two.

   When the teeth of the dog have accomplished their eruption,
they cease to grow and their roots remain firmly fixed in the
alveolar cavities.   In the first dentition only the premolars are
found ; the post-molars appear as permanent teeth.

The molars in the dog do not form simple arches as in the other animals, but each arch represents a double curve; so that the two arches on either jaw make the form of a lyre, with its base backward and its apex forward, the widest portion of which is on a line between the last premolar and the first molar. The form or curve of the arches varies considerably with the different races of dogs, and increases greatly in those breeds which, by domestication, have had their heads and maxillæ shortened. These changes may almost approach the nature of a deformity, and the last premolar and first molar may be crowded en-

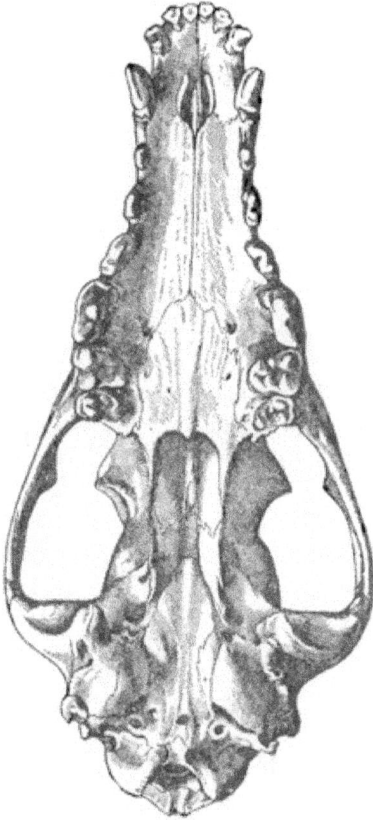

FIG. 177.
Skull of a greyhound, under surface.

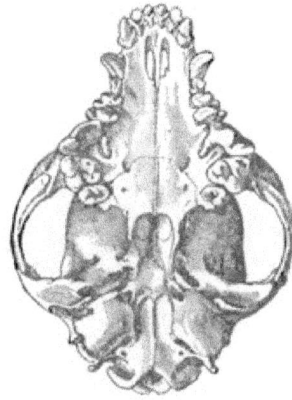

FIG. 178.
Skull of a pug, under surface.

tirely out of the line of the other teeth, or may be absent, as is sometimes seen in the pug or fancy bull-dogs.

Dr. James A. Waugh informs me that in the Mexican hairless dogs the teeth are generally irregular in number and relative position in the jaws. They have no tubercles or trefoils on the table surface of the incisor teeth. They have no canine teeth, or tushes, in either jaw. He has met with some half-breds that had canine teeth in the upper or the lower jaw, but found no cases with those teeth in both jaws. The molars vary in number and relative position in different animals. The development of the adult dentition is much slower and later than in other classes of dogs.

One Mexican hairless dog, aged 2 years, had five superior incisors, no canine teeth, two superior molars on each side, one large molar and one very small molar behind the large one; two inferior incisors, no canine teeth, two inferior molars on each side, one large molar and one very small molar behind the large one. Total number of teeth, 15.

A Mexican hairless bitch, aged 2 years, had four superior incisors in front and one on right side midway between the front teeth and the molars, no canine teeth, two superior molars on each side, one small molar and one large molar behind the smaller one; seven inferior incisors in front, and two on each side midway between the front teeth and the molars, no canine teeth, two inferior molars on each side, one large molar and one small molar behind the large one. Total number of teeth, 24.

DETERMINATION OF AGE BY THE TEETH.

From the evidences furnished by the teeth the age is divided into three periods:—

1. Eruption of the temporary teeth.
2. Eruption of the permanent teeth.
3. Wearing of the permanent teeth.

ERUPTION OF THE TEMPORARY TEETH.

At Birth.—Puppies may be born with all of their temporary teeth, but if the teeth have not appeared the eruption commences at once by the incisors and tusks of the upper jaw, and the entire dentition is effected within the first three weeks at the longest. (Fig. 179.)

Two to Four Months.—A month before the eruption of the permanent teeth the temporary pinchers and often the intermediate teeth in both jaws become loose and fall out; at this time the points of the permanent teeth can be felt by pressing on the gums. (Fig. 180.)

ERUPTION OF THE PERMANENT TEETH.

The permanent teeth replace those of first dentition several months earlier in the large races of dog than in the terrier varieties. Medium-sized dogs, setters, etc., make the change later than the large dogs and earlier than the terriers.

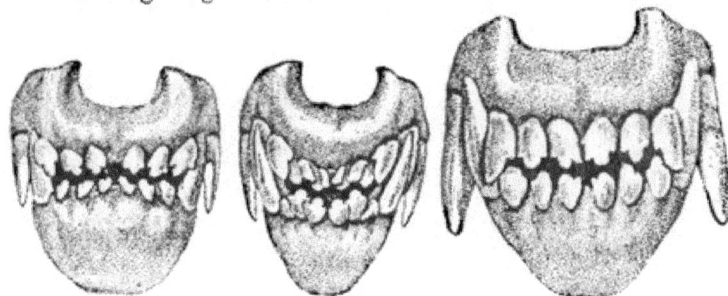

FIG. 179.   FIG. 180.   FIG. 181.

FIG. 179.—Teeth of puppy two months old.

FIG. 180.—Teeth of dog four months old; the incisors are loose and ready to fall out.

FIG. 181.—Teeth of dog one year old, showing the incisors and tushes fresh and unworn.

Five to Eight Months.—According to the race of dog, the permanent teeth appear rapidly in the following order: pinchers, intermediate, corners, and tusks. The incisors do not come out obliquely, as in the herbivora. It is worthy of note that, while the herbivora do not have all of their permanent dentition until they have arrived at their full development of body, the carnivora have theirs long before they have attained their growth and before the consolidation of their bones.

One year.—For the first few months after their appearance there is no alteration in the teeth; at one year they are pure white and the tubercles on the incisors are intact. (Fig. 181.)

Fifteen Months.—From fifteen months on, commences the wearing away of the teeth. This is first seen on the inferior pinchers, while the tusks and other teeth remain fresh and white in color.

Eighteen Months to Two Years.—At about this time the tubercles of the inferior pinchers become worn down and those of the other incisors commence to be used.   (Fig. 182.)

Two and One-half Years to Three Years.—After two and a half years the tubercles of the inferior intermediate teeth are worn away, the incisors of the upper jaw show signs of use, and all of the teeth commence to lose their fresh, white appearance and become yellowish and discolored.   (Fig. 183.)

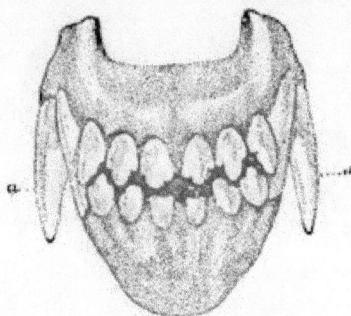

FIG. 182.

Teeth of dog eighteen months old, showing use of the inferior pinchers. The tusks are still fresh and white.

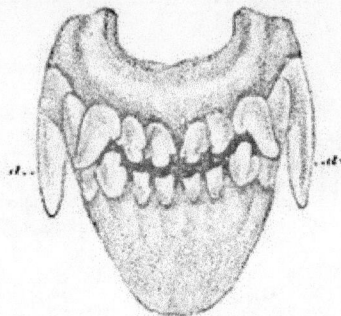

FIG. 183.

Teeth of dog two and a half years old, showing use of inferior pinchers and intermediate teeth.

Three and One-half Years to Four Years.—The pinchers of the upper jaw become worn down and are followed by the wearing of the superior intermediate teeth.

The discoloration of the teeth increases and the tusks become yellow and dirty in appearance.

After this the leveled and broken teeth give no indication of the age.   A Great Dane, which for twelve years had only been fed at the table on tid-bits, and a poodle of thirteen years, had teeth as fresh as those of an ordinary dog at four or five; and bull-terriers and others at four or five may have the incisors nearly worn away.

The leveling of the teeth of the dog consists of the wearing away of the middle tubercle to the level of the lateral tubercles and the wearing of the enamel from these. As we have just seen, the leveling commences with the inferior pinchers, is followed by that of the inferior intermediate teeth, and does not commence on the superior incisors until it is complete on the inferior ones; with the superior incisors it commences with the pinchers and follows with the intermediate teeth; the tusks do not commence to wear until the use of the incisors is marked. The leveling of the teeth may, however, take place much more rapidly or more slowly than has been indicated above, and may even be very irregular in the teeth which it first affects, according to the nature of the food upon which the animal is nourished.

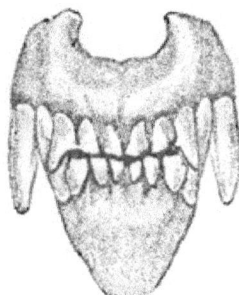

Fig. 184.

Teeth of dog over three years old, showing use of all of the incisors; the tushes are becoming yellow.

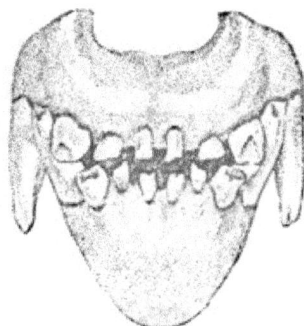

Fig. 185.

Teeth of an old dog; all of the incisors and tushes are much used and worn.

Dogs fed on meat and those which have bones to knaw use their teeth more rapidly than those which are fed on soups, broken bread, and vegetables. Dogs which are pugnacious, and those which are taught to swing on ropes or straps and to fetch objects, such as sticks, stones, etc., use their teeth rapidly and are apt to have their tusks much worn. These dogs also frequently have irregular mouths, from broken teeth. From these causes it sometimes happens, especially with the larger dogs, that the teeth are so altered after three years that no intelligible deduction can be drawn from the appearance of the mouth.

DETERMINATION OF AGE FROM OTHER SIGNS.

The young dog is born with its eyelids closed; these do not open until the ninth or tenth and sometimes as late as the fifteenth day.

Young dogs have heads which are large in proportion to their size, and in which the cranium is large in proportion to the size of the face.

Old dogs become gray around the nose, eyes, and forehead; their noses become larger, and the skin over the whole face becomes wrinkled. The lips become everted and show the red mucous membrane along their irregular borders. The skin, at the ordinary points of pressure in decubitus, especially in large dogs, becomes denuded of hair and develops an epithelial growth which is almost horny in character; so that in old dogs we find callous points at the elbow, hock, salient points of the pelvis, and sometimes on the shoulder. They are apt to develop eczema and other skin troubles of a chronic nature. Frequent epithelial tumors appear over the surface of the body. In the old male growths on the penis and in the old female tumors in the mammary glands are not rare. Deafness, without apparent lesion in the ears, and cataracts, with a congested condition of the conjunctiva, occur. Paralysis of the hind-legs, from no evident cause, may end the old dog's life.

THE age of man has always been a subject of the most profound philosophical study; for, from the time of the philosopher's stone to the researches in alchemy of the Middle Ages for the source of life everlasting, and to the elixirs of life of the quack medicines of to-day, it has always been the desire of a large number of mankind to outlive its time. From a practical point of view, the question of the age of man is one of sentiment. The periods of *babyhood, youth, adult* life, and *old age* are definitely marked, and depend not altogether upon the number of years which the animal has lived, but to a greater degree upon the mental development, which, in a precocious youth, makes a man of him, and in an adult, who should have been in his prime, but has abused his natural surroundings, makes prematurely an old man.

The mother, as a matter of sentiment, looks to the dentition of her baby for an evidence of its development, and the teeth furnish medico-legal proof of the age of certain subjects; so that the dentition of man, as in the other animals, is of first importance as an indication of age, during a certain period, until adult life.

DENTITION OF MAN.

Dental formula
$$\begin{cases} \text{Temporary,} & \cdot \quad \frac{2 \cdot 1 \cdot 2}{2 \cdot 1 \cdot 2} = 20 \\ \\ \text{Permanent,} & \cdot \quad \frac{2 \cdot 1 \cdot 2 \cdot 3}{2 \cdot 1 \cdot 2 \cdot 3} = 32 \end{cases}$$

Man has thirty-two teeth,—four incisors in each jaw (8), two tushes in each jaw (4), and five molars (which are subdivided into two premolars and three molars) in each arch of both jaws (20). The arch of the teeth, unlike in the other animals, is continuous, no interspace existing between the incisors and the tushes, nor between the latter and the molars.

The upper arch is slightly larger than the lower, and the corresponding teeth of each jaw are not directly opposite each other, but come in contact with two opposite teeth, so that, if one is lost, the opposing tooth still finds a resisting surface, and has not the tendency to displacement found in the herbivora under similar circumstances.

The incisors, four in each jaw, are known as the *central* and *lateral* incisors; they have a single root and a sharp, wedge-shaped crown, divided from the former by a distinct neck.

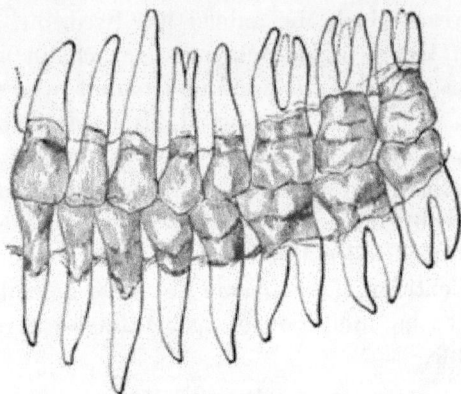

Fig. 186.
Upper and lower teeth of one-half of the dental arches in each jaw; two incisors, one tush, two premolars, three molars, in each.

The tushes, called *cuspids* (*cuspis*, a spear), from their pointed crowns, are two in number in each jaw. These teeth are also called *canine* teeth; those of the upper jaw are called *eye* teeth, and those of the lower jaw *stomach* teeth.

The premolars, two in each side of jaw, are called *bicuspids* (*bicuspidati*, two spear-points), from the two eminences on their crowns.

The molars, six in each jaw, are known as the *first, second,* and *third,* or as the *six-year* molars, *twelve-year* molars, and the *wisdom* teeth (*dens sapientia*), from the fact that the two first appear at six and twelve years, and the last not until adult age, when wisdom is supposed to come.

The incisors are double wedge-shape,—widest at the cutting edge, thickest at the neck, slightly concave on the inner and slightly convex on the outer surface. The superior or upper are larger than the inferior or lower incisors, and the central incisors are wider than the lateral. In the case of the inferior incisors, the laterals are wider than the centrals. The incisors, both superior and inferior, have each but one root. The use of these teeth in eating is to " bite " or cut off a portion of food, operating on the same principle as a pair of scissors. The cuspids are less concave on the inner surface and more convex on the outer surface than the incisors, and terminate in a point. They have but one root, but this is longer and stronger than that of any other tooth in the mouth. The superior cuspids are larger and have longer roots than the inferior. The use of the cuspids is to seize and tear or lacerate obstinate substances preparatory to mastication. It is an interesting fact that these teeth are more prominent in proportion as the animal approaches the purely carnivorous or flesh-eating class, and are never found in an animal having horns. The bicuspids are smaller than the cuspids, convex on both outer and inner surfaces, and flattened on the sides. Their long diameter is across the jaw. The roots are conical. The inferior have but one root; the superior sometimes a single root, often deeply grooved, and sometimes entirely divided,—bifid. The molars have large grinding surfaces, divided by grooves into cusps or points. The crowns of the inferior are larger than those of the superior. In each jaw they decrease in size from before backward. The superior first and second molars have, as a rule, three roots each. The inferior first and second molars have each two roots, which are often deeply grooved, and sometimes bifid. The third molars or wisdom-teeth of both jaws have but a single root each, though this is sometimes divided into three in the upper jaw and two in the lower. The inferior third molar is larger than the superior.

The deciduous or temporary teeth are much smaller than the permanent, although the roots are generally larger and longer, in proportion to the size of the crowns, than those of the adult set. There are only twenty teeth in the deciduous set,—four incisors, two cuspids, and four molars in each jaw. Figs. 190 and 191 represent the temporary teeth,—superior and inferior. There are no bicuspids and no third molars or wisdom-teeth in the temporary set. Four bicuspids and two wisdom-teeth in each jaw,—twelve in all,—which are not in the deciduous or temporary set, make the permanent set to consist of thirty-two teeth.

The surfaces of the teeth which are toward the lips are called "labial;" toward the cheeks, "buccal;" toward the tongue, on the lower jaw, "lingual;" and toward the roof of the mouth, on the upper jaw, "palatal." The surfaces next to each other are called "approximal;" those looking toward the

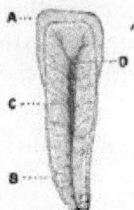 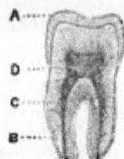 

Fig. 187.   Fig. 188.   Fig. 189.

centre, "mesial;" and those looking from the centre, "distal." The parts of the six front teeth of both jaws which come in contact with each other are called the "cutting edges," and the broad surfaces of the bicuspids and molars which are brought in contact in the act of masticating are called the "grinding or articulating surfaces."

The structure of the teeth is approximately that of the tush-tooth of the horse or of the molars of the dog, consisting of a central pulp covered with dentine, which makes up the bulk of the tooth, and which is again covered with enamel. The cement in man covers the *root only*, or at most only slightly overlaps the enamel.

Figs. 187 and 188 represent a central incisor and a molar, split vertically so as to show their various parts. A is the

cutting edge or grinding surface, covered, as is the entire
crown, with enamel; B. the cementum covering the roots; C,
the dentine; D, the pulp-cavity.   Fig. 189 shows a transverse
section of a molar, of natural size, in which 1 is the dentine; 2,
the enamel; 3, the pulp-cavity.

The pulp is *not* replaced by dentine as the animal becomes
adult, but remains vascular and with a nerve-supply, and when
exposed by disease of the tooth causes the toothache, which is
not possible in the adult herbivora.

DETERMINATION OF AGE BY THE TEETH.

TEMPORARY DENTITION.

The baby is born without teeth and depends entirely upon
the mother's milk or artificially prepared food which does not
require mastication until the fifth month, when the eruption,
called *cutting teeth*, so well known in every household, commences with the central incisors.   The lower teeth of each set
usually precede the upper by several weeks to two or three
months.

The teeth usually appear in pairs, but often that of one side
precedes the other by some weeks, and in rare cases several
pairs may appear at once.   The eruption of the teeth in the
baby is frequently attended with much pain, and sometimes by
grave and serious disturbance of the digestive and nervous
systems, terminating even in death.   These complications are
rare in other animals, but occasionally occur in young dogs,
which may have gastro-enteritis or convulsions during the eruption of the permanent teeth, and we have already noted a case
of convulsions in a two-year-old colt.   An intercurrent disease
during dentition may delay or alter its processes.

An increased flow of saliva, hot and dry skin, constipation
and diarrhœa, eruptions of the skin and ulcers in the mouth
are frequent attendant symptoms of the appearance of the teeth.
Difficult dentition is accompanied by itching of the nose, twitch-

ing of the muscles, dilatation of the pupils, and evidences of
suffering and fever.

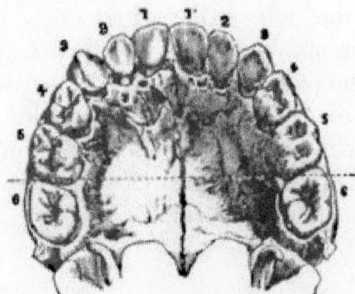

Fig. 190.
Dentition of child six years old, upper jaw.

The eruption of the temporary teeth is shown in the following table:—

Two central incisors (No. 1) between the fifth and eighth months.
Two lateral incisors (No. 2) between the seventh and tenth months.
Two canines (No. 3) between the twelfth and sixteenth months.
Two first molars (No. 4) between the fourteenth and twentieth
    months.
Two second molars (No. 5) between the twentieth and thirty-sixth
    months.

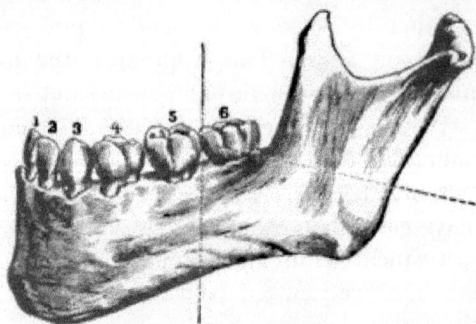

Fig. 191.
Dentition of child six years old, lower jaw.

The teeth marked "6," in Figs. 190 and 191, do not belong
to the temporary set.

PERMANENT DENTITION.

The permanent teeth are well developed and pressing on the roots of the temporary ones before the latter are shed.

Fig. 192 illustrates the jaws of a child about six years of age, in which the relations of the two sets of teeth are shown,—the temporary teeth still in position, and the sixth-year molars just erupting. The permanent incisors, both upper and lower,—which, after the sixth-year molars, are the next to erupt,—are the most advanced, both as to completeness of form and as to

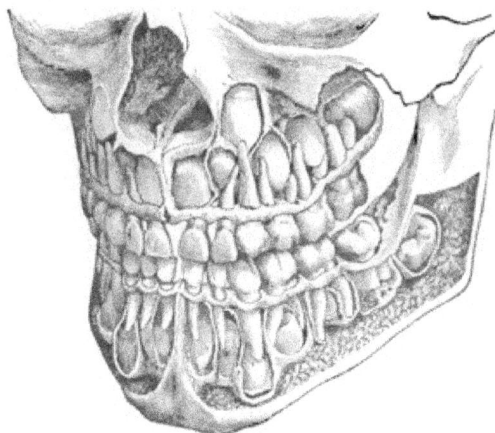

FIG. 192.

position. The canines, which are not due until the child is between eleven and thirteen years of age, are not nearly so complete in form nor so far advanced toward eruption.

The permanent teeth appear in the following order :—

First molars,	5 to 6 years.
Central incisors,	6 to 8 years.
Lateral incisors,	7 to 9 years.
First bicuspids,	9 to 10 years.
Second bicuspids,	10 to 11 years.
Canines,	11 to 13 years.
Second molars,	12 to 14 years.
Wisdom-teeth,	17 to 21 years.

14

Fig. 193 illustrates an upper permanent or adult set of teeth. These teeth are known by the same names as the corresponding teeth of the lower jaw.

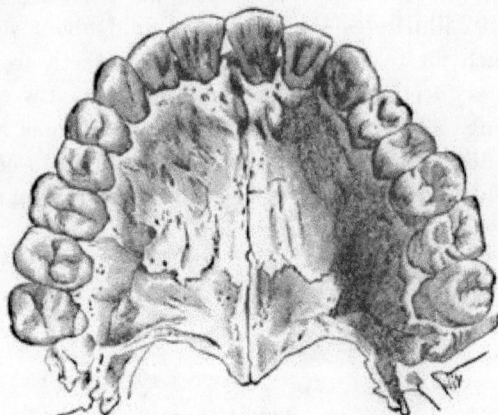

FIG. 193.

Fig. 194 is a side view of the adult lower jaw, in which No. 1 represents the central incisor; No. 2, the lateral incisor; No. 3, the canine; No. 4, the first bicuspid; No. 5, the second bicuspid; No. 6, the first (sixth-year) molar; No. 7, the second (twelfth-year) molar; and No. 8, the third molar, or wisdom-tooth

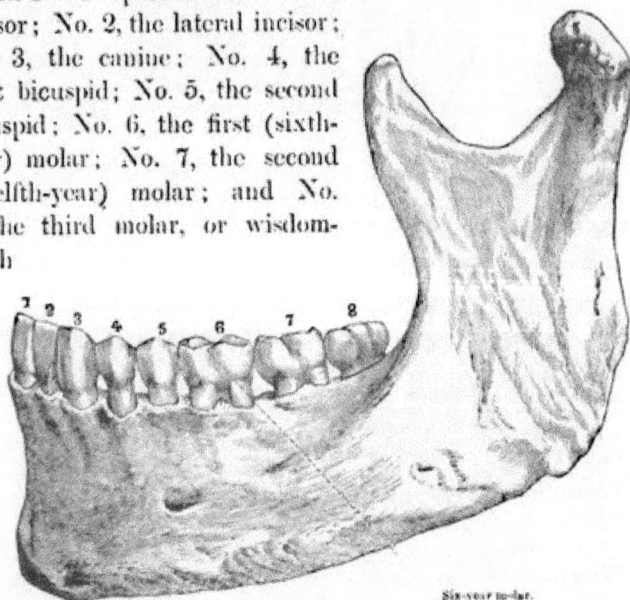

Six-year molar.

FIG. 194.

BIBLIOGRAPHY.

GOUBAUX AND BARRIER.—De l'Exterieur du cheval. Par MM. Armand Goubaux et Gustave Barrier, deuxiéme edition. Paris: Asselin et Houzeau, 1890.

CHAUVEAU.—Traité d'Anatomie comparée des Animaux Domestiques. Par A. Chauveau. 3me edition, de S. Arloing, Paris. J. B. Baillière et Fils, 1879.

GIRARD.—Traité de l'Age du Cheval. Par feu N.-F. Girard. 3me edition, etc., augmentée de l'age du boeuf, du mouton, du chien, et du cochon, par F. Girard. Paris: Bechat Jeune.

BRACY CLARK.—On the Knowledge of the Age of the Horse by his Teeth. London, 1826.

MAYHEW.—The Horse's Mouth, etc. By Edward Mayhew.

H. BOULEY. Ages. Nouveau Dictionnaire pratique de Médecine de chirurgie et d'hygiene vétérinaires. Paris, 1856.

REYNAL.—Age. Ibid.

SIMONDS.—The Age of the Ox, Sheep, and Pig, etc. By James Beart Simonds. London, 1854.

OLOF SCHWARTZKOPFF.—Modern Feeding of Pigs, and its Influence upon the Formation of the Skull and Dentition.

J. W. WHITE, M.D., D.D.S.—The Mouth and the Teeth. P. Blakiston, Son, & Co., Philadelphia, 1891.

G. CUVIER.—Leçons d'anatomie comparée.

TOUSSAINT.—De l'age des chevaux de course, au point de vue de la doctrine de la precocité. Recueil de Méd. Vét., 1875, fol. 992.

——Traité de la connoissance exterieure de cheval, avec un examen analytique de toutes les fourberies des Maquignons. Turin, 1769.

F. LOCOQ.—Traité de l'Exterieur du Cheval et des principaux animaux domestiques. 5me edition. P. Asselin, Paris, 1876.

ANDRÉ SANSON.—Traité de Zootechnie. Paris, 1874.

A. RAILLIET.—Eléments de Zoölogie Médicale et Agricole. Paris, 1886.

CARLO RUINI.—Anatomia del Cavallo, Gasparo Bindoni. Venitia, MDXCIX.

(211)

212

BIBLIOGRAPHY.

FRANCESCO LIBERATI ROMANO.—La perfettione del Cavallo.   Roma,1639.

PESSINA.—Sul modo de conoscere dai denti l'età dei Cavalli.   Milan, 1831.

CARLO LESSONA.—Compendio d'Ippiatria.   Torino, 1846.

YOUATT.—History, Treatment, and Diseases of the Horse.

W. H. CLARKE.—Horse's Teeth.   Turf, Field, and Farm.   New York, 1883.

LIAUTARD.—How to tell the Age of the Domestic Animals.   Jenkins, New York.

ELLENBERGER AND BAUM.—Anatomie des Hundes.   By Dr. W. Ellenberger and Dr. H. Baum.   Paul Parey, Berlin, 1891.

CH. CORNEVIN.—Traité de Zootechnie Générale.   J. B. Baillière et Fils. Paris, 1891.

CORNEVIN ET LESBRE.—Charactères ostéologiques différentiels de la chévre et du mouton.   Journal de Méd. Vét. et de Zootechnie, September, 1891.

CRUZEL.

DAUBENTON.—Histoire Naturelle, etc.

ERICK-VIBORG.

JAMES A. WAUGH.

INDEX.

Abnormal triangularity, 127
Absence of incisors, 127
  of superior incisors, ox, 158
Absorption of bone, 15
Aëropinic, 146
Age, juvenile, 1
  adult, 1
  senile, 2
Age of ass, 157
  of dog, 194
  of goat, 175
  of hinny, 157
  of hog, 184
  of horse, 9
  of man, 203
  of mule, 157
  of ox, 158
  of sheep, 175
Aged, 48
Alveolar periosteum, 22, 141
Angle of incidence of incisors, 29
Arch of incisors, 29
Artificial cup, 155
  irregularities, 151
Ass, age of, 157
  supernumerary tush tooth, 126
Augmentation of teeth, 123

Barren mares, 31
Berkshire, 187
Biangular table, 21
Bicuspids, 205
Birth of hog, 187
  of horse, 53, 55
Bishoping, 155
Brachygnathism, 123

Calf, incisors of, 164
Canaliculi, 26
Canine teeth, horse, 31
Caps of molars, 35, 36, 37
Cartilaginous beds, ox, 161
  cushion, ox, 158
Carlo Ruini, error, 33
Cauterization, 153, 154, 155
Cavity of dental pulp, 12
Cement, molars, 42
  excessive deposits, 25
  use of, 25
Cementoma, 140
Characters furnished by teeth, 51
Collection of skulls, 186
Contagious diseases, 43
  fevers, 16
Convergence of incisors, 31
Convulsions, 15, 43

Cores of horns, 172
Corner teeth, horse, 16
Cribbing, 146
Crown, incisors, horse, 17
  molars, 37
Cup, artificial, 155
  double, 127
  of teeth, 12, 17
Cushion of upper jaw, ox, 160
Cuspids, 205
Cutting teeth, 207

Date of birth, 44
  common-bred, 44
  thoroughbreds, 44
Deciduous molars, horse, 33
  teeth, 11
Deficiency of length, 140
  in length of crown, 141
Dehorned, 171
Dental cavity, 12
  cup, 12, 17
  formula, 8
  of dog, 194
  of goat, 175
  of hog, 184
  of horse, 9
  of man, 203
  of ox, 158
  of sheep, 175
Dental star, 19, 26, 93
  change of position, 28
  ox, 161
Dentine, 22
  molars, 41
Dentists, equine, 152
Dentition of dog, 194
  of goat, 175
  of hog, 184
    table of, 189
  of horse, 9
  of man, 203
  of ox, 158
  of sheep, 176
Depth of cups, 19, 128
Determination of age, dog, 198, 202
  hog, 186
  horse, 48
    by skin, 49
    by knots in tail, 50
    by teeth, 50
  man, 207
  ox, by horns, 171
    by teeth, 164
  sheep, 182
Development, molars, 40

(213)

Direction of incisors, 29
    ox, 160
    of plane of contact, 29
    in regard to median line, 30
Disease of jaw, 141
    of molars, 144
Dishorned, 171
Divergence of incisors, 30
Diverticula, 40
Dog, age of, 194
    dental formula, 194
    dentition, 194
    determination of age, 198, 202
    incisors, 195
    irregular teeth, 201
    molars, 196
    permanent teeth, eruption, 199
    temporary teeth, eruption, 198
    tusks, 195
Dogs, Mexican, 197
Double cup, 127
    rows of incisors, 123
Dressed mouth, 156
Duration of life of horse, 46
Durham, 174

Eburnated substance, 26
Effect of surroundings on age, 2
Eight months, 54, 62, 63
    temporary molars, 33
Eighteen months, sheep, 179
Enamel, 25
    germ, 21
    of the cup, 22
    molars, 41
    peripheral, 22
Epidermic covering of horns, 172-182
Epochs of age, 1
Equine dentists, 152
Error, Carlo Ruini, 33
Eruption, hasty, 43
    oblique, 152
    of teeth, hog, 187
    of horse, 43
    of incisors, horse, 43, 53
    of temporary teeth, goat, 177
    of permanent incisors, ox, 165
    of temporary teeth, ox, 164
    of sheep, 177
    of tusks, 41
    of molars, 44
    table of, horse, 45
    tardy, 43
Evolution of skull, hog, 191
Ewe, 180
Excess of length of crown, 142
Excessive length, 128
    of jaws, 136
    of single teeth, 140
    of teeth, 136

Feeding, irregular, 143
Fifth period, ox, 171
    eleven years, 171

Fifth period, ox, twelve years, 171
First dentition, molars, horse, 33
Fissure of incisors, 128
    in mules, 128
Five months, 53, 60, 61
    years, ox, 168
Fleur-de-lis, 195
Fœtal teeth, 11
Form of horns, 174
Four months, 53, 59
    years, sheep, 181
Fourteen years, irregular, 132, 133
Fourth period, 92
    horse, 92
      six years, 92, 94, 95
      seven years, 92, 96, 97
      eight years, 92, 98, 99
    ox, 168
      five years, 168
      six years, 168
      seven years, 169
      eight years, 170
      nine years, 170
      ten years, 170
Fraudulent acts, 137, 153
    horns, ox, 173
Fundamental substance, 26

Germ of the enamel, 21
    ivory, 21
Germ-follicle, 21
Goat, age of, 175
    horns, 182
    incisors of, 176
    irregular incisors, 182
    molars, 177
    old, 182
    structure of horns, 182
Greyhound, skull, 197
Grooves of horns, 173
Growth of horns, 173, 182
    sheep, 182
Gutters, molars, 34
Gyps, 155

Hardness of bone, thoroughbred, 51
    enamel, 25
    teeth, 128
Hasty eruption of teeth, 43
Hasten age, 152
Hinny, age of, 157
Histology, 26
Hog, birth, 187
    dental formula, 184
    dentition, 184
    determination of age, 186
    eruption of teeth, 187
    incisors, 184
    milk dentition, 189
    molars, 185
    permanent teeth, 188
    precocity, 186
    table of dentition, 189
    temporary teeth, 187

Hog, tushes, 185
Horizontality of incisive arches, 30
Horns, 182
  epidermic covering, 172-182
  form of, 174
  goat, 182
  growth of, 173
  ox, 171
  scraped, 173
  shape of, 172
  sheep, 182
  show-cattle, 173
Hot iron to gums, 153
Hungarian ox, 174
Hypertrophy of molar, 145

Incidence, angle of, 29
Incisors, absent, 127
  irregular, goats, 182
  of calf, 159, 164
  of dog, 195
  of eruption, 152
  of hog, 184
  of horse, 11
    first dentition, 11
    permanent, 16
  of man, 204
  of ox, 159
  of sheep, 176
  structure, 160
  united, 127
Inferior molars, horse, 40
Infundibula, molars, 40
Interlobular spaces of Czermak, 27
Intermediate teeth, horse, 16
Introduction, 1
Irregular feeding, 142
Irregularities of the dental system, 122
  artificial, 151
  bishoping, 123
  depth of dental cup, 122
  diminution, 122
  dog, 201
  excess of size of jaw, 122
  excess or fault of use, 123
  fault of length of jaw, 122
  filing, 123
  fissure of the dental cup, 122
  form of the cup, 122
  form of the incisors, 122
  fourteen years, 132, 133
  fraudulent alterations, 123
  goat, 182
  in number, 123
  marks produced by cribbing, 123
  nine years, 130, 131
  sheep, 180
  size of the cavity of cup, 122
  uniting of two incisors, 122
Is — years, 51
  — years off, 51
Ivory, 22
  germ, 21
  molars, 41

January, date of birth of thorough-
  breds, 44
Jaw, diseased, 144
Jersey, 174
Jura ox, 174

Knots in tail, 50

Lamb, 180
Leveling, ox, 160
Longevity of horse, 47

Man, age of, 203
  dentition, 203
  dental formula, 203
  incisors, 204
  table of eruption of temporary
    teeth, 207
  table of eruption of permanent
    teeth, 207
  temporary dentition, 207
  tushes, 204
Mares' tusks, 31
Mark of mouth, 100
Marks of cribbing, 148-151
Maxillary sinus, 39
May, date of birth of common-bred, 44
Merino ram, 183
Mexican dogs, 197
Milk dentition, hog, 189
  molars, horse, 33
  teeth, horse, 11
  remaining, 125
Montenegrin ram, 183
Moon-blindness, 44
Mule, age of, 157
Mulley, 171
Molars, 201
  cups, 35, 36, 37
  development, 40
  diseased, 144
  diverticula, 40
  dog, 196
  first dentition, horse, 33
  goat, 177
  hog, 185
  horse, 33
  hypertrophy, 145
  inferior, horse, 40
  infundibula, 40
  man, 205
  milk, horse, 33
  ox, 161
  permanent dentition, 35
  second dentition, 35
  sheep, 177
  structure, 40
  subzygomatic, 126
  superior, 36
  supernumerary, 126

Nine years, irregular, 130, 131
Number of teeth, horse, 9

Oblique eruption, incisors, 152

Obliquity of incisive arches, 30
Off years, 51
Ogive, 29
Old dogs, 202
   English 3, 37, 40
   goat, 182
   horses, 46
   sheep, 182
One month, 53, 57
   week, 53, 56
   year, sheep, 179
Oval table, 19
Ox, age of, 158
   dental formula, 158
     star, 161
   determination of age by horns, 171
     by teeth, 164
   dentition of, 158
   direction of incisors, 160
   eruption of permanent incisors, 165
     of temporary teeth, 164
   molars of, 161
   permanent molars, 161
   rudimentary molars, 158
   table of dentition, 158
   table of eruption of teeth, 163
   temporary incisors, 160
     molars, 161
   wearing of temporary teeth, 164

Papilla of the enamel, 22
   of the ivory, 22
   molars, 41
Papillæ, 21
Parrot-mouth, 137, 138
   reserved, 139
Past mark of mouth, 100
Period, 189
Periods, growth, 2–4
   stationary, 2–5
   decline, 2–5
   of age, 1
   of horse, 53
     eruption of incisors of first dentition, 53
     leveling of temporary incisors, 64
     falling out of the temporary teeth, 74
     appearance of the permanent teeth, 74
     leveling of permanent teeth, 92
     successive forms, tables of permanent teeth, 100
   of ox, 162
     eruption of temporary teeth, 162
     wearing of the temporary teeth, 162
     eruption of the permanent teeth, 162
     leveling of permanent teeth, 162
     wearing away of the crowns, 162
   of eruption, horse, 51
     first period, 53

Periods of eruption, horse, second period, 64
   third period, 74
   fourth period, 92
   fifth period, 100
Periodic ophthalmia, 43
Peripheral enamel, 22
Permanent dentition, hog, 189
   man, 208
   incisors, horse, 16
     ox, eruption, 165
   molars, horse, 35
     ox, 161
   teeth, dog, 198
   teeth, hog, 188
Pinchers, horse, 16
Poland China, 187
Polled, 171
   hangus, 171
Polling by Hindoos, 171
Post-molars, 36
   horse, 33
Precocity of hog, 186
   of horse, 47
   of sheep, 175
Premolars, 36, 204
   horse, 33
   man, 205
Primitive diameter, table of incisors, 29
Principles of examination for age, 48
Prognathism, 123, 140
Pulp-cavity, 12, 19

Queue d'Hirondelle, 181

Ram, 180
Relative angle of incisors, 29
Removing temporary incisors, 152
Résumé, direction of incisors, 31
Reversed parrot-mouth, 139
Rings of horns, 173
Rising — years, 51
Roots, molars, 34
   permanent molars, horse, 39
Rounded table, 19
Rudimentary molars, ox, 158

Second dentition molars, horse, 35
   period, horse, 64
   one year, 64, 66, 67
   sixteen months, 64, 68, 69
   twenty months, 64, 70, 71
   two years, 64, 72, 73
Sections of incisors, antero-posterior, 26
   horse, 19
Sheep, age of, 175
   eruption of incisors, 177
   eruption of molars, 177
   horns of, 182
   incisors, 176
   irregularities of teeth, 180
   molars, 177
   old, 182
   precocity, 175

Sheep, structure of horns, 183
Sinus, maxillary, 39
Six months, 54
Six-year molars, 204
Skull of greyhound, 197
    of hog, evolution, 191
    of pug, 197
    of Texas boar, 192
    of Yorkshire sow, 192
Star, dental, 19, 26
Structure of horns, goat, 182
        sheep, 182
    of molars, 40
    of incisors, 21
        ox, 160
    of tushes, 32
Sub-zygomatic molars, 126
Superior molars, 36
Supernumerary intermediate teeth, 123
    molars, 126
    pinchers, 123
    teeth, 125
    tush teeth, 126
Superstition, 15, 31
Swallow-tails, 181

Table, 21, 27
    biangular, 21
    oval, 19
    rounded, 19
    transverse, 19
    triangular, 21
Table, change of form, 28
Table of dentition of hog, 189
    of ox, 158
    of eruption of teeth, ox, 163
    of incisors, ox, 160
    of teeth, 17, 27
Tangents, 29
Tardy eruption of teeth, 43
The teeth, 7
Teeth of rodents, 184
Temporary dentition, man, 207
    incisors, ox, 160
        removed, 152
    molars, ox, 161
    teeth, dog, 198
        hog, 187
        horse, 11
        ox, eruption, 164

Temporary teeth, ox, wearing, 164
    tusks, 32
Ten months, 54, 62, 63
Texas boar, skull, 192
Three months, 53, 58
    years, sheep, 180
Third period, horse, 74
    two and a half years, 74
    rising three years, 74, 78, 79
    three years, 74
    three years off, 75, 80, 81
    rising four years, 75, 82, 83
    four years, 75, 84, 85
    four years off, 75, 86, 87
    rising five years, 76, 88, 89
    five years, 76, 90, 91
    five years off, 76
Transverse incisors, 125
    table, 19
Triangular, abnormal, 127
    table, 21
Tushes, horse, 31
    ox, 158
    supernumerary, 126
Tusks of dog, 195
    of hog, 185
    of horse, 31
Twelve-year molars, 204
Two years, sheep, 179

Uniting of two incisors, 127
Use of cement, 25
    teeth, 7-9
Used, 48

Virgin tooth, 12

Wearing of molars, 42
    of teeth, 136
        dog, 198
    of temporary teeth, ox, 164
Wind-sucking, 146
Wisdom-teeth, 204
Wolf-teeth, 33, 158
    ox, 158

—years off, 51
Yorkshire sow, skull, 192
Young animals, 3

THE JOURNAL OF
Comparative Medicine
AND Veterinary Archives.

EDITED BY

W. A. CONKLIN, Ph.D., D.V.S.,
Director of the Zoological Gardens, New York;

AND

R. S. HUIDEKOPER, M.D., VETERINARIAN,
Professor of Sanitary Medicine and Veterinary Jurisprudence, American Veterinary College, New York.

A MONTHLY JOURNAL

which offers to the VETERINARY PROFESSION, to PHYSICIANS interested in Comparative Medicine, and to the intelligent AGRICULTURIST, STOCK-BREEDER and HORSEMAN, the most complete record of the discoveries in Comparative Medicine and of the advance in Veterinary matters that can be found in any one publication in the English language.

All matters of general interest in animal industry, the breeding, care of, shoeing, use of, and commerce in animals, are carefully reported. These allied subjects are as useful to the HORSEMAN and AGRICULTURIST as they are necessary to the VETERINARIAN.

During the last year articles were published from the pens of Berns, Coates, Bryden, Butler, Clement, Curtice, Fleming, Hoskins, Liautard, Lyford, Meyers, Morris, Paquin, Peters, Salmon, Schwartzkopff, Schufeldt, Sibley, J. Bland Sutton, Welch, Williams, and many others, the most prominent Veterinarians in the world.

The readers are assured that Volume XIII for 1892 will surpass the previous editions in the matter contained, pertaining to Comparative Medicine, Clinical Records, Laboratory Research, and Animal Industry. Illustrations are used whenever they can aid in rendering the writings more clear.

Full reports of the Veterinary Schools of the United States and Canada, and the proceedings of the local, State and United States Veterinary Medical Associations are published promptly.

The JOURNAL is the organ of no school, but is the representative of the VETERINARY PROFESSION OF THE UNITED STATES AND CANADA.

PUBLISHED MONTHLY, AT THREE DOLLARS A YEAR IN ADVANCE.

ADDRESS ALL COMMUNICATIONS AND PAYMENTS TO

JOURNAL COMPARATIVE MEDICINE AND VETERINARY ARCHIVES,
S. W. COR. 42D STREET AND PARK AVENUE, NEW YORK.

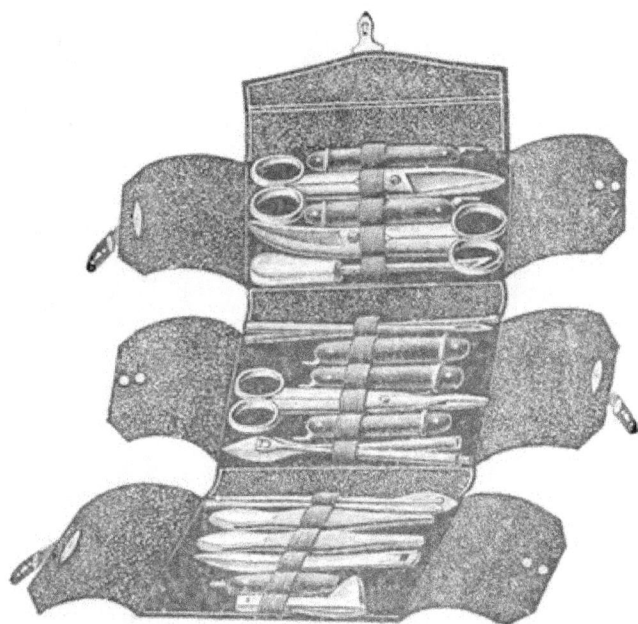

SEPTEMBER, 1891.

CATALOGUE

OF THE

☞ In place of repeated revisions, Supplements to this Catalogue will be issued at intervals of about every three months. These supplements will be regularly mailed to all those who will favor us with their name and address.

MEDICAL

PUBLICATIONS

OF

F. A. DAVIS,

MEDICAL PUBLISHER AND BOOKSELLER,

1231 FILBERT STREET,

PHILADELPHIA, PA.

BRANCH OFFICES:

NEW YORK CITY—117 W. Forty-Second St.

CHICAGO—24 Lakeside Building, 214-220 S. Clark Street.

ATLANTA—69 Old Capitol.

LONDON, ENG.—40 Berners Street, Oxford Street, W.

Order from Nearest Office.

For Sale by all Booksellers.

SPECIAL NOTICE.

PRICES of books, as given in our catalogues and circulars, include full prepayment of postage, freight, or express charges. Customers in Canada and Mexico must pay the cost of duty, in addition, at point of destination.

N. B.—Remittances should be made by Express Money-Order, Post-Office Money-Order, Registered Letter, or Draft on New York City, Philadelphia, Boston, or Chicago.

We do not hold ourselves responsible for books sent by mail; to insure safe arrival of books sent to distant parts, the package should be registered. Charges for registering (at purchaser's expense), ten cents for every four pounds, or less.

INDEX TO CATALOGUE.

PAGE

Annual of the Universal Medical
Sciences 27, 28, 29

Anatomy.

Practical Anatomy—Boenning   4
Structure of the Central Nervous Sys-
tem—Edinger   8
Charts of the Nervo-Vascular System—
Price and Eagleton 17
Synopsis of Human Anatomy—Young . . 25

Bacteriology.

Bacteriological Diagnosis—Eisenberg . .   8

Clinical Charts.

Improved Clinical Charts—Bashore . . .   3

Consumption.

Consumption: How to Prevent it, etc.—
Davis   7

Domestic Hygiene, etc.

The Daughter: Her Health, Education,
and Wedlock—Capp   5
Consumption: How to Prevent it, etc.—
Davis   7
Plain Talks on Avoided Subjects—
Guernsey   9
Heredity, Health, and Personal Beauty—
Shoemaker 21

Electricity.

Practical Electricity in Medicine and
Surgery—Liebig and Rohé 12
Electricity in the Diseases of Women—
Massey 13

Fever.

Fever: its Pathology and Treatment—
Hare 10
Hay Fever—Sajous 19

Gynecology.

Lessons in Gynecology—Goodell   9

Heart, Lungs, Kidneys, etc.

Diseases of the Heart, Lungs, and
Kidneys—Davis   7
Diseases of the Heart and Circulation in
Children—Keating and Edwards . . . 12
Diabetes: its Cause, Symptoms, and
Treatment—Purdy 17

Hygiene.

American Resorts—James 11
Text-Book of Hygiene—Rohé 18

Materia Medica and Thera-
peutics.

Hand-Book of Materia Medica, Phar-
macy, and Therapeutics—Bowen . . .   4
Ointments and Oleates—Shoemaker . . 21
Materia Medica and Therapeutics—Shoe-
maker 22
International Pocket Medical Formulary
—Witherstine 26

Miscellaneous.

PAGE

Book on the Physician Himself—Cathell .   5
Oxygen—Demarquay and Wallian   7
Record-Book of Medical Examinations
for Life Insurance—Keating 11
The Medical Bulletin, Monthly 13
Physician's Interpreter 15
Circumcision—Remondino 18
Medical Symbolism—Sozinskey 23
International Pocket Medical Formulary
—Witherstine 26
The Chinese: Medical, Political, and
Social—Coltman 31
A, B, C of the Swedish System of Educa-
tional Gymnastics—Nissen 32
Lectures on Auto-Intoxication—Bouchard 32

Nervous System, Spine, etc.

Spinal Concussion—Clevenger   6
Structure of the Central Nervous System
—Edinger   8
Epilepsy: its Pathology and Treatment—
Hare 10
Lectures on Nervous Diseases—Ranney . 30

Obstetrics.

Childbed: its Management: Diseases and
Their Treatment—Manton 12
Eclampsia—Michener and others 15
Obstetric Synopsis—Stewart 24

Pharmacology.

Abstracts of Pharmacology—Wheeler . . 25

Physiognomy.

Practical and Scientific Physiognomy—
Stanton 30

Physiology.

Physiology of the Domestic Animals—
Smith 23

Surgery and Surgical Operations.

Circumcision—Remondino 18
Principles of Surgery—Senn 20

Swedish Movement and Massage.

Swedish Movement and Massage Treatment—Nissen 15

Throat and Nose.

Journal of Laryngology and Rhinology . 11
Hay Fever—Sajous 19
Diphtheria, Croup, etc.—Sanne 19
Lectures on the Diseases of the Nose and
Throat—Sajous 31

Venereal Diseases.

Syphilis: To-day and in Antiquity—Buret   4
Neuroses of the Genito-Urinary System
in the Male—Ultzmann 24

Veterinary.

Age of the Domestic Animals—Huide-
koper 32
Physiology of the Domestic Animals—
Smith 23

Visiting-Lists and Account-
Books.

Medical Bulletin Visiting-List or Physicians' Call-Record 14
Physicians' All-Requisite Account-Book . 16

BASHORE'S IMPROVED CLINICAL CHART.

For the SEPARATE PLOTTING of TEMPERATURE, PULSE, and RESPIRATION.

Designed for the Convenient, Accurate, and Permanent Daily Recording of Cases in Hospital and Private Practice.

By HARVEY B. BASHORE, M.D.

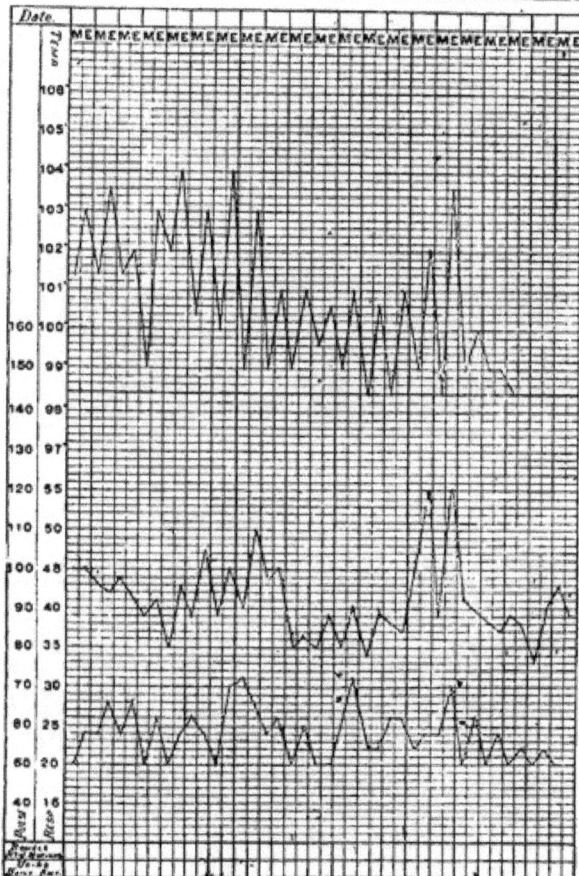

50 Charts, in Tablet Form.   **Size, 8 x 12 inches.**

Price, in the United States and Canada, Post-paid, 50 Cents, Net; Great Britain, 2s. 8d.; France, 3 fr. 60.

The above diagram is a little more than one-fifth (⅕) the actual size of the chart and shows the method of plotting, the upper curve being the Temperature, the middle the Pulse, and the lower the Respiration. By this method a full record of each can easily be kept with but one color Ink.

It is so arranged that all practitioners will find it an invaluable aid in the treatment of their patients.

On the back of each chart will be found ample space conveniently arranged for recording "Clinical History and Symptoms" and "Treatment."

By its use the physician will secure such a complete record of his cases as will enable him to review them at any time. Thus he will always have at hand a source of individual improvement and benefit in the practice of his profession, the value of which can hardly be overestimated.

(F. A. DAVIS, Medical Publisher, Philadelphia, Pa., U.S.A.)

CLEVENGER

Spinal Concussion.

SURGICALLY CONSIDERED AS A CAUSE OF SPINAL INJURY, AND NEURO-
LOGICALLY RESTRICTED TO A CERTAIN SYMPTOM GROUP, FOR WHICH
IS SUGGESTED THE DESIGNATION ERICHSEN'S DISEASE,
AS ONE FORM OF THE TRAUMATIC NEUROSES.

By S. V. CLEVENGER, M.D., Consulting Physician Reese and Alexian
Hospitals; Late Pathologist County Insane Asylum, Chicago; Member
of numerous American Scientific and Medical Societies; Collaborator
American Naturalist, Alienist and Neurologist, Journal of Neurology
and Psychiatry, Journal of Nervous and Mental Diseases; author of
"Comparative Physiology and Psychology," "Artistic Anatomy," etc.

This work is the outcome of five years' special study and experience
in legal circles, clinics, hospital and private practice, in addition to
twenty years' labor as a scientific student, writer, and teacher.

The literature of Spinal Concussion has been increasing of late years
to an unwieldy shape for the general student, and Dr. Clevenger has in
this work arranged and reviewed all that has been done by observers
since the days of Erichsen and those who preceded him.

There are abundant illustrations, particularly for Electro-diagnosis,
and to enable a clear comprehension of the anatomical and pathological
relations.

The Chapters are: I. Historical Introduction; II. Erichsen on
Spinal Concussion; III. Page on Injuries of the Spine and Spinal Cord;
IV. Recent Discussions of Spinal Concussion; V. Oppenheim on
Traumatic Neuroses; VI. Illustrative Cases from Original and all other
Sources; VII. Traumatic Insanity; VIII. The Spinal Column; IX.
Symptoms; X. Diagnosis; XI. Pathology; XII. Treatment; XIII.
Medico-legal Considerations.

Other special features consist in a description of modern methods
of diagnosis by Electricity, a discussion of the controversy concerning
hysteria, and the author's original pathological view that the lesion is
one involving the spinal sympathetic nervous system. In this latter
respect entirely new ground is taken, and the diversity of opinion concerning the functional and organic nature of the disease is afforded a
basis for reconciliation.

Every Physician and Lawyer should own this work.

In one handsome Royal Octavo Volume of nearly 400 pages, with
thirty Wood-Engravings.

**Price, post-paid, in United States and Canada, $2.50, net; in Great
Britain, 14s.; in France, 15 fr.**

The reader will find in this book the best
discussion and summary of the facts on this
topic, which will make it very valuable to
every physician. For the specialist it is a
text-book that will be often consulted.—*The
Journal of Inebriety.*

The work comes fully up to the demand,
and the law and medical library, to be complete, cannot be without it.—*Southern Medical
Record.*

This work really does, if we may be per-
mitted to use a trite and hackneyed expression, "fill a long-felt want." The subject is
treated in all its bearings; electro-diagnosis
receives a large share of attention, and the
chapter devoted to illustrative cases will be
found to possess especial importance. The
author has some original views on pathology.
—*Medical Weekly Review.*

DAVIS

Consumption: How to Prevent it, and How to Live with it.

Its NATURE, CAUSES, PREVENTION, AND THE MODE OF LIFE, CLIMATE, EXERCISE, FOOD AND CLOTHING NECESSARY FOR ITS CURE.

By N. S. DAVIS, JR., A.M., M.D., Professor of Principles and Practice of Medicine in Chicago Medical College; Physician to Mercy Hospital; Member of the American Medical Association, Illinois State Medical Society, etc., etc. 12mo. IN PRESS.

DAVIS

Diseases of the Heart, Lungs, and Kidneys.

By N. S. DAVIS, JR., A.M., M.D., Professor of Principles and Practice of Medicine in Chicago Medical College; Physician to Mercy Hospital; Member of the American Medical Association, Illinois State Medical Society, etc., etc. In one neat 12mo volume. *No.   in the Physicians' and Students' Ready-Reference Series.* IN PREPARATION.

DEMARQUAY

On Oxygen. A Practical Investigation of the Clinical and Therapeutic Value of the Gases in Medical and Surgical Practice,

WITH ESPECIAL REFERENCE TO THE VALUE AND AVAILABILITY OF OXYGEN, NITROGEN, HYDROGEN, AND NITROGEN MONOXIDE.

By J. N. DEMARQUAY, Surgeon to the Municipal Hospital, Paris, and of the Council of State; Member of the Imperial Society of Surgery; Correspondent of the Academies of Belgium, Turin, Munich, etc ; Officer of the Legion of Honor, Chevalier of the Orders of Isabella-the-Catholic and of the Conception, of Portugal, etc.   Translated, with notes, additions, and omissions, by SAMUEL S. WALLIAN, A.M., M.D., Member of the American Medical Association; Ex-President of the Medical Association of Northern New York; Member of the New York County Medical Society, etc.

In one handsome Octavo Volume of 316 pages, printed on fine paper, in the best style of the printer's art, and illustrated with 21 Wood-Cuts.

Price, post-paid, in United States, Cloth, $2.00, net; Half-Russia, $3.00, net.   In Canada (duty paid), Cloth, $2.20, net; Half-Russia, $3.30, net.   In Great Britain, Cloth, 11s. 6d.; Half-Russia, 17s. 6d.   In France, Cloth, 12 fr. 40; Half-Russia, 18 fr. 60.

For some years past there has been a growing demand for something more satisfactory and more practical in the way of literature on the subject of what has, by common consent, come to be termed "Oxygen Therapeutics." On all sides professional men of standing and ability are turning their attention to the use of the gaseous elements about us as remedies in disease, as well as sustainers in health. In prosecuting their inquiries, the first hindrance has been the want of any reliable, or in any degree satisfactory, literature on the subject.

This work, translated in the main from the French of Professor Demarquay, contains also a very full account of recent English, German, and American experiences, prepared by Dr. Samuel S. Wallian, of New York, whose experience in this field antedates that of any other American writer on the subject.

This is a handsome volume of 300 pages, in large print, on good paper, and nicely illustrated. Although nominally pleading for the use of oxygen inhalations, the author shows in a philosophical manner how much greater good physicians might do if they more fully appreciated the value of fresh air exercise and water, especially in diseases of the lungs, kidneys, and skin. We commend its perusal to our readers.—*The Canada Medical Record.*

The book should be widely read, for to many it will bring the addition of a new weapon to their therapeutic armament.—*Northwestern Lancet.*

Altogether the book is a valuable one, which will be found of service to the busy practitioner who wishes to keep abreast of the improvements in therapeutics.—*Medical News.*

EISENBERG

Bacteriological Diagnosis.

TABULAR AIDS FOR USE IN PRACTICAL WORK.

By JAMES EISENBERG, Ph.D., M.D., Vienna. Translated and augmented, with the permission of the author, from the latest German Edition, by NORVAL H. PIERCE, M.D., Surgeon to the Out-Door Department of Michael Reese Hospital; Assistant to Surgical Clinic, College of Physicians and Surgeons, Chicago, Ill.

This book is a novelty in Bacteriological Science. It is arranged in a tabular form in which are given the specific characteristics of the various well-established bacteria, so that the worker may, at a glance, inform himself as to the identity of a given organism. They then serve the same function to the Bacteriologist as does the "Chemical Analysis Chart" to the chemist, and the one will be found as essential as the other.

THE GREATEST care has been taken to bring the work up to the present aspect of Bacteriology.

In one Octavo volume, handsomely bound in Cloth. READY SOON.

Price, post-paid, in the United States and Canada, $1.50, net; in Great Britain, 8s. 6d.; in France, 9 fr. 35.

EDINGER

Twelve Lectures on the Structure of the Central Nervous System.

FOR PHYSICIANS AND STUDENTS.

By DR. LUDWIG EDINGER, Frankfort-on-the-Main. Second Revised Edition. With 133 Illustrations. Translated by WILLIS HALL VITTUM, M.D., St. Paul, Minn. Edited by C. EUGENE RIGGS, A.M., M.D., Professor of Mental and Nervous Diseases, University of Minnesota; Member of the American Neurological Association.

The illustrations are exactly the same as those used in the latest German edition (with the German names translated into English), and are very satisfactory to the Physician and Student using the book.

The work is complete in one Royal Octavo volume of about 250 pages, bound in Extra Cloth.

Price in United States and Canada, post-paid, $1.75, net; Great Britain, 10s.; France, 12 fr. 20.

One of the most instructive and valuable works on the minute anatomy of the human brain extant. It is written in the form of lectures, profusely illustrated, and in clear language. The book is worthy of the highest encomiums, and will, undoubtedly, command a large sale.—*The Pacific Record of Medicine and Surgery.*

Since the first works on anatomy, up to the present day, no work has appeared on the subject of the general and minute anatomy of the central nervous system so complete and exhaustive as this work of Dr. Ludwig Edinger. Being himself an original worker, and having the benefits of such masters as Stilling, Weigert, Gerlach, Meynert, and others, he has succeeded in transforming the mazy wilderness of nerve fibres and cells into a district of well-marked pathways and centres, and by so doing has made a pleasure out of an anatomical bugbear.—*The Southern Medical Record.*

Every point is clearly dwelt upon in the text, and where description alone might leave a subject obscure clever drawings and diagrams are introduced to render misconception of the author's meaning impossible. The book is eminently practical. It unravels the intricate entanglement of different tracts and paths in a way that no other book has done so explicitly or so concisely.—*Northwestern Lancet.*

(8)

GOODELL

Lessons in Gynecology.

By WILLIAM GOODELL, A.M., M.D., etc., Professor of Clinical Gynecology in the University of Pennsylvania.

This exceedingly valuable work, from one of the most eminent specialists and teachers in gynecology in the United States, is now offered to the profession in a much more complete condition than either of the previous editions. It embraces all the more important diseases and the principal operations in the field of gynecology, and brings to bear upon them all the extensive practical experience and wide reading of the author. It is an indispensable guide to every practitioner who has to do with the diseases peculiar to women. Third Edition. With 112 illustrations. Thoroughly revised and greatly enlarged. One volume, large octavo, 578 pages.

Price, in United States and Canada, Cloth, $5.00; Full Sheep, $6.00. Discount, 20 per cent., making it, net, Cloth, $4.00; Sheep, $4.80. Postage, 27 cents extra. Great Britain, Cloth, 22s. 6d.; Sheep, 28s., post-paid. France, 30 fr. 80.

It is too good a book to have been allowed to remain out of print, and it has unquestionably been missed. The author has revised the work with special care, adding to each lesson such fresh matter as the progress in the art rendered necessary, and he has enlarged it by the insertion of six new lessons. This edition will, without question, be as eagerly sought for as were its predecessors.—*American Journal of Obstetrics.*

His literary style is peculiarly charming. There is a directness and simplicity about it which is easier to admire than to copy. His chain of plain words and almost blunt expressions, his familiar comparison and homely illustrations, make his writings, like his lectures, unusually entertaining. The substance of his teachings we regard as equally excellent.—*Philadelphia Medical and Surgical Reporter.*

Extended mention of the contents of the book is unnecessary; suffice it to say that every important disease found in the female sex is taken up and discussed in a common-sense kind of a way. We wish every physician in America could read and carry out the suggestions of the chapter on "the sexual relations as causes of uterine disorders—conjugal onanism and kindred sins." The department treating of nervous counterfeits of uterine diseases is a most valuable one.—*Kansas City Medical Index.*

GUERNSEY

Plain Talks on Avoided Subjects.

By HENRY N. GUERNSEY, M.D., formerly Professor of Materia Medica and Institutes in the Hahnemann Medical College of Philadelphia; author of Guernsey's "Obstetrics," including the Disorders Peculiar to Women and Young Children; Lectures on Materia Medica, etc. The following Table of Contents shows the scope of the book:

CONTENTS.—Chapter I. Introductory. II. The Infant. III. Childhood. IV. Adolescence of the Male. V. Adolescence of the Female. VI. Marriage: The Husband. VII. The Wife. VIII. Husband and Wife. IX. To the Unfortunate. X. Origin of the Sex. In one neat 16mo volume, bound in Extra Cloth.

Price, post-paid, in the United States and Canada, $1.00; Great Britain, 6s.; France, 6 fr. 20.

HARE

Epilepsy: Its Pathology and Treatment.

BEING AN ESSAY TO WHICH WAS AWARDED A PRIZE OF FOUR THOUSAND FRANCS BY THE ACADEMIE ROYALE DE MEDECINE DE BELGIQUE, DECEMBER 31, 1889.

By HOBART AMORY HARE, M.D. (Univ. of Penna.), B.Sc., Professor of Materia Medica and Therapeutics in the Jefferson Medical College, Phila.; Physician to St. Agnes' Hospital and to the Children's Dispensary of the Children's Hospital; Laureate of the Royal Academy of Medicine in Belgium, of the Medical Society of London, etc.; Member of the Association of American Physicians.

No. 7 in the Physicians' and Students' Ready-Reference Series. 12mo. 228 pages. Neatly bound in Dark-blue Cloth.

Price, post-paid, in United States and Canada, $1.25, net; in Great Britain, 6s. 6d.; in France, 7 fr. 75.

It is representative of the most advanced views of the profession, and the subject is pruned of the vast amount of superstition and nonsense that generally obtains in connection with epilepsy.—*Medical Age.*

Every physician who would get at the gist of all that is worth knowing on epilepsy, and who would avoid useless research among the mass of literary nonsense which pervades all medical libraries, should get this work."—*The Sanitarian.*

It contains all that is known of the pathology of this strange disorder, a clear discussion of the diagnosis from allied neuroses, and the very latest therapeutic measures for relief.

It is remarkable for its clearness, brevity, and beauty of style. It is, so far as the reviewer knows, altogether the best essay ever written upon this important subject.—*Kansas City Medical Index.*

The task of preparing the work must have been most laborious, but we think that Dr. Hare will be repaid for his efforts by a wide appreciation of the work by the profession; for the book will be instructive to those who have not kept abreast with the recent literature upon this subject. Indeed, the work is a sort of Dictionary of epilepsy—a reference guide-book upon the subject.—*Alienist and Neurologist.*

HARE

Fever: Its Pathology and Treatment.

BEING THE BOYLSTON PRIZE ESSAY OF HARVARD UNIVERSITY FOR 1890. CONTAINING DIRECTIONS AND THE LATEST INFORMATION CONCERNING THE USE OF THE SO-CALLED ANTIPYRETICS IN FEVER AND PAIN.

By HOBART AMORY HARE, M.D. (Univ. of Penna.), B.Sc., Professor of Materia Medica and Therapeutics in the Jefferson Medical College, Phila.; Physician to St. Agnes' Hospital and to the Children's Dispensary of the Children's Hospital; Laureate of the Royal Academy of Medicine in Belgium, of the Medical Society of London, etc.; Member of the Association of American Physicians.

No. 10 in the Physicians' and Students' Ready-Reference Series. 12mo. Neatly bound in Dark-blue Cloth.

Illustrated with more than 25 new plates of tracings of various fever cases, showing beautifully and accurately the action of the Antipyretics. The work also contains 35 carefully prepared statistical tables of 249 cases showing the untoward effects of the antipyretics.

Price, post-paid, in the United States and Canada, $1.25, net; in Great Britain, 6s. 6d.; in France, 7 fr. 75.

As is usual with this author, the subject is thoroughly handled, and much experimental and clinical evidence, both from the author's experience and that of others, is adduced in support of the view taken.—*New York Medical Abstract.*

The author has done an able piece of work in showing the facts as far as they are known concerning the action of antipyrin, antifebrin, phenacetin, thallin, and salicylic acid. The reader will certainly find the work one of the most interesting of its excellent group, the *Physicians' and Students' Ready-Reference Series.*—*The Postmortem Medical Review.*

Such books as the present one are of service to the student, the scientific therapeutist, and the general practitioner alike, for much can be found of real value in Dr. Hare's book, with the additional advantage that it is up to the latest researches upon the subject.—*University Medical Magazine.*

JAMES

American Resorts. With Notes upon their Climate.

By BUSHROD W. JAMES, A.M., M.D., Member of the American Public Health Association, and the Academy of Natural Sciences, Philadelphia; the Society of Alaskan Natural History and Ethnology, Sitka, Alaska, etc. With a translation from the German, by MR. S. KAUFFMANN, of those chapters of "Die Klimate der Erde" written by DR. A. Woeikof, of St. Petersburg, Russia, that relate to North and South America and the Islands and Oceans contiguous thereto.

This is a unique and valuable work, and useful to physicians in all parts of the country. We mention a few of the merits it possesses: *First*, List of all the Health Resorts of the country, arranged according to their climate. *Second*, Contains just the information needed by tourists, invalids, and those who visit summer or winter resorts. *Third*, The latest and best large railroad map for reference. *Fourth*, It indicates the climate each one should select for health. *Fifth*, The author has traveled extensively, and most of his suggestions are practical in reference to localities. In one Octavo volume. Handsomely bound in Cloth. Nearly 300 pages.

Price, post-paid, in the United States and Canada, $2.00, net;
Great Britain, 11s. 6d.; France, 12 fr. 10.

Taken altogether, this is by far the most complete exposition of the subject of resorts that has yet been put forth, and it is one that every physician must needs possess intelligent information upon.—*Buffalo Med. & Surg. Jour.*

The special chapter on the therapeutics of climate . . . is excellent for its precautionary suggestions in the selection of climates and local conditions, with reference to known pathological indications and constitutional predispositions.—*The Sanitarian.*

The book before us is a very comprehensive volume, giving all necessary information concerning climate, temperature, humidity, sunshine, and indeed everything necessary to be stated for the benefit of the physician or invalid seeking a health resort in the United States.—*Southern Clinic.*

Journal of Laryngology and Rhinology.

ISSUED ON THE FIRST OF EACH MONTH.

Edited by Dr. Norris Wolfenden, of London, and Dr. John Macintyre, of Glasgow, with the active aid and co-operation of Drs. Dundas Grant, Barclay J. Baron, Hunter Mackenzie, and Sir Morell Mackenzie. Besides those specialists in Europe and America who have so ably assisted in the collaboration of the Journal, a number of new correspondents have undertaken to assist the editors in keeping the Journal up to date, and furnishing it with matters of interest. Amongst these are: Drs. Sajous, of Philadelphia; Middlemass Hunt, of Liverpool; Mellow, of Rio Janeiro; Sedziak, of Warsaw; Draispul, of St. Petersburg, etc. Drs. Michael, Joal, Holger, Mygind, Prof. Massei, and Dr. Valerius Idelson will still collaborate the literature of their respective countries.

Price, 13s. or $3.00 per annum (inclusive of Postage). For single copies, however, a charge of 1s. 3d. (30 cents) will be made. Sample Copy, 25 Cents.

KEATING

Record-Book of Medical Examinations

FOR LIFE INSURANCE.

Designed by JOHN M. KEATING, M.D.

This record-book is small, neat, and complete, and embraces all the principal points that are required by the different companies. It is made in two sizes, viz.: No. 1, covering one hundred (100) examinations, and No. 2, covering two hundred (200) examinations. The size of the book is 7 x 3½ inches, and can be conveniently carried in the pocket.

	U. S. and Canada.	Great Britain.	France.
No. 1. For 100 Examinations, in Cloth, - -	$ .50 Net	3s. 6d.	3 fr. 60
No. 2. For 200 Examinations, in Full Leather, with Side Flap, - - - -	1.00 "	6s.	6 fr. 20

KEATING and EDWARDS

Diseases of the Heart and Circulation.

IN INFANCY AND ADOLESCENCE. WITH AN APPENDIX ENTITLED "CLINICAL STUDIES ON THE PULSE IN CHILDHOOD."

By JOHN M. KEATING, M.D., Obstetrician to the Philadelphia Hospital, and Lecturer on Diseases of Women and Children; Surgeon to the Maternity Hospital; Physician to St. Joseph's Hospital; Fellow of the College of Physicians of Philadelphia, etc.; and WILLIAM A. EDWARDS, M.D., Instructor in Clinical Medicine and Physician to the Medical Dispensary in the University of Pennsylvania; Physician to St. Joseph's Hospital; Fellow of the College of Physicians; formerly Assistant Pathologist to the Philadelphia Hospital, etc.

Illustrated by Photographs and Wood-Engravings. About 225 pages. Octavo. Bound in Cloth.

Price, post-paid, in the United States and Canada, $1.50, net; in Great Britain, 8s. 6d.; in France, 9 fr. 35.

Drs. Keating and Edwards have produced a work that will give material aid to every doctor in his practice among children. The style of the book is graphic and pleasing, the diagnostic points are explicit and exact, and the therapeutical resources include the novelties of medicine as well as the old and tried agents.—*Pittsburgh Med. Review.*

It is not a mere compilation, but a systematic treatise, and bears evidence of considerable labor and observation on the part of the authors. Two fine photographs of dissections exhibit mitral stenosis and mitral regurgitation; there are also a number of wood-cuts.—*Cleveland Medical Gazette.*

LIEBIG and ROHÉ

Practical Electricity in Medicine and Surgery.

By G. A. LIEBIG, JR., PH.D., Assistant in Electricity, Johns Hopkins University; Lecturer on Medical Electricity, College of Physicians and Surgeons, Baltimore; Member of the American Institute of Electrical Engineers, etc.; and GEORGE H. ROHÉ, M.D., Professor of Obstetrics and Hygiene, College of Physicians and Surgeons, Baltimore; Visiting Physician to Bay View and City Hospitals; Director of the Maryland Maternité; Associate Editor "Annual of the Universal Medical Sciences," etc.

Profusely Illustrated by Wood-Engravings and Original Diagrams, and published in one handsome Royal Octavo volume of 383 pages, bound in Extra Cloth.

The constantly increasing demand for this work attests its thorough reliability and its popularity with the profession, and points to the fact that it is already THE standard work on this very important subject. The part on Physical Electricity, written by Dr. Liebig, one of the recognized authorities on the science in the United States, treats fully such topics of interest as Storage Batteries, Dynamos, the Electric Light, and the Principles and Practice of Electrical Measurement in their Relations to Medical Practice. Professor Rohé, who writes on Electro-Therapeutics, discusses at length the recent developments of Electricity in the treatment of stricture, enlarged prostate, uterine fibroids, pelvic cellulitis, and other diseases of the male and female genito-urinary organs. The applications of Electricity in dermatology, as well as in the diseases of the nervous system, are also fully considered.

Price, post-paid, in the United States and Canada, $2.00, net; in Great Britain, 11s. 6d.; France, 12 fr. 40.

Any physician, especially if he be a beginner in electro-therapeutics, will be well repaid by a careful study of this work by Liebig and Rohé. For a work on a special subject the price is low, and no one can give a good excuse for remaining in ignorance of so important a subject as electricity in medicine.—*Toledo Medical and Surgical Reporter.*

The entire work is thoroughly scientific and practical, and is really what the authors have aimed to produce, "a trustworthy guide to the application of electricity in the practice of medicine and Surgery."—*New York Medical Times.*

In its perusal, with each succeeding page, we have been more and more impressed with the fact that here, at last, we have a treatise on electricity in medicine and surgery which amply fulfills its purpose, and which is sure of general adoption by reason of its thorough excellence and superiority to other works intended to cover the same field.—*Pharmaceutical Era.*

After carefully looking over this work, we incline to the belief that the intelligent physician who is familiar with the general subject will be greatly interested and profited.—*American Lancet.*

MASSEY

Electricity in the Diseases of Women.

WITH SPECIAL REFERENCE TO THE APPLICATION OF STRONG CURRENTS.

By G. BETTON MASSEY, M.D., Physician to the Gynæcological Department of the Howard Hospital; late Electro-therapeutist to the Philadelphia Orthopædic Hospital and Infirmary for Nervous Diseases; Member of the American Neurological Association, of the Philadelphia Neurological Society, of the Franklin Institute, etc. SECOND EDITION. Revised and Enlarged. With New and Original Wood-Engravings. Handsomely bound in Dark-Blue Cloth. 240 pages. 12mo. *No. 5 in the Physicians' and Students' Ready-Reference Series.*

This work is presented to the profession as the most complete treatise yet issued on the electrical treatment of the diseases of women, and is destined to fill the increasing demand for clear and practical instruction in the handling and use of strong currents after the recent methods first advocated by Apostoli. The whole subject is treated from the present stand-point of electric science *with new and original illustrations*, the thorough studies of the author and his wide clinical experience rendering him an authority upon electricity itself and its therapeutic applications. The author has enhanced the practical value of the work by including *the exact details* of treatment and results in a number of cases taken from his private and hospital practice.

Price, post-paid, in the United States and Canada, $1.50, net; in Great Britain, 8s. 6d.; in France, 9 fr. 35.

A new edition of this practical manual attests the utility of its existence and the recognition of its merit. The directions are simple, easy to follow and to put into practice; the ground is well covered, and nothing is assumed, the entire book being the record of experience.—*Journal of Nervous and Mental Diseases.*

It is only a few months since we noticed the first edition of this little book; and it is only necessary to add now that we consider it the best treatise on this subject we have seen, and

that the improvements introduced into this edition make it more valuable still.—*Boston Medical and Surgical Journ.*

The style is clear, but condensed. Useless details are omitted, the reports of cases being pruned of all irrelevant material. The book is an exceedingly valuable one, and represents an amount of study and experience which is only appreciated after a careful reading.—*Medical Record.*

MANTON

Childbed; Its Management; Diseases and Their Treatment.

By WALTER P. MANTON, M.D., Visiting Physician to the Detroit Woman's Hospital; Consulting Gynæcologist to the Eastern Michigan Asylum; President of the Detroit Gynæcological Society; Fellow of the American Society of Obstetricians and Gynæcologists, and of the British Gynæcological Society; Member of Michigan State Medical Society, etc. In one neat 12mo volume. *No. in the Physicians' and Students' Ready-Reference Series.* IN PREPARATION.

Medical Bulletin.

A MONTHLY JOURNAL OF MEDICINE AND SURGERY.

Edited by JOHN V. SHOEMAKER, A.M., M.D. Bright, original, and readable. Articles by the best practical writers procurable. Every article as brief as is consistent with the preservation of its scientific value. Therapeutic Notes by the leaders of the medical profession throughout the world. These, and many other unique features, help to keep THE MEDICAL BULLETIN in its present position as the leading low-price Medical Monthly of the world. Subscribe now.

**TERMS: $1.00 a year in advance in United States, Canada, and Mexico.
Foreign Subscription Terms: England, 5s.; France, 6 fr.; Germany,
6 marks; Japan, 1 yen; Australia, 5s.; Holland, 3 florins.**

The Medical Bulletin Visiting-List or Physicians' Call Record.

ARRANGED UPON AN ORIGINAL AND CONVENIENT MONTHLY AND WEEKLY
PLAN FOR THE DAILY RECORDING OF PROFESSIONAL VISITS.

Frequent Rewriting of Names Unnecessary.

THIS Visiting-List is arranged so that the names of patients need be written but ONCE a month instead of FOUR times a month, as in the old-style lists. By means of a new feature, a simple device consisting of STUB OR HALF LEAVES IN THE FORM OF INSERTS, the first week's visits are recorded in the usual way, and the second week's visits are begun by simply turning over the half-leaf without the necessity of rewriting the patients' names. This very easily understood process is repeated until the month is ended and the record has been kept complete in every detail of VISIT, CHARGE, CREDIT, etc., and the labor and time of entering and transferring names at least THREE times in the month has been saved. There are no intricate rulings ; not the least amount of time can be lost in comprehending the plan, for it is acquired at a glance.

THE THREE DIFFERENT STYLES MADE.

The **No. 1 Style** of this List provides space for the DAILY record of seventy different names each month for a year ; for physicians who prefer a List that will accommodate a larger practice we have made a **No. 2 Style**, which provides space for the daily record of 105 different names each month for a year, and for physicians who may prefer a Pocket Record-Book of less thickness than either of these styles we have made a **No. 3 Style**, in which "The Blanks for the Recording of Visits in" have been made into removable sections. These sections are very thin, and are made up so as to answer in full the demand of the largest practice, each section providing ample space for the DAILY RECORD OF 210 DIFFERENT NAMES for two months ; or 105 different names daily each month for four months ; or seventy different names daily each month for six months. Six sets of these sections go with each copy of No. 3 STYLE.

SPECIAL FEATURES NOT FOUND IN ANY OTHER LIST.

In this No. 3 STYLE the PRINTED MATTER, and such matter as the BLANK FORMS FOR ADDRESSES OF PATIENTS, Obstetric Record, Vaccination Record, Cash Account, Birth and Death Records, etc., are fastened permanently in the back of the book, thus reducing its thickness. The addition of one of these removable sections does not increase the thickness more than an eighth of an inch. This brings the book into such a small compass that no one can object to it on account of its thickness, as its bulk is VERY MUCH LESS than that of any visiting-list ever published. Every physician will at once understand that as soon as a section is full it can be taken out, filed away, and another inserted without the least inconvenience or trouble. *Extra or additional sections will be furnished at any time for 15 cents each or $1.75 per dozen.* This Visiting-List contains calendars, valuable miscellaneous data, important tables, and other useful printed matter usually placed in Physicians' Visiting-Lists.

Physicians of many years' standing and with large practices pronounce it THE BEST LIST THEY HAVE EVER SEEN. It is handsomely bound in fine, strong leather, with flap, including a pocket for loose memoranda, etc., and is furnished with a Dixon lead-pencil of excellent quality and finish. It is compact and convenient for carrying in the pocket. Size, 4 x 6¾ inches.

IN THREE STYLES

	NET PRICES.
No. 1. Regular size, to accommodate 70 patients daily each month for one year,	$1.25
No. 2. Large size, to accommodate 105 patients daily each month for one year,	$1.50
No. 3. In which the "Blanks for Recording Visits in" are in removable sections,	$1.75
Special Edition for Great Britain, without printed matter,	4s. 6d.

● *N. B.—The Recording of Visits in this List may be Commenced at any time during the Year.*

MICHENER

Hand-Book of Eclampsia; OR, NOTES AND CASES OF PUERPERAL CONVULSIONS.

By E. MICHENER, M.D.; J. H. STUBBS, M.D.; R. B. EWING, M.D.; B. THOMPSON, M.D.; S. STEBBINS, M.D. 16mo. Cloth.

Price, 60 cents, net; in Great Britain, 4s. 6d.; France, 4 fr. 20.

NISSEN

A MANUAL OF INSTRUCTION FOR GIVING

Swedish Movement and Massage Treatment

By PROF. HARTVIG NISSEN, late Director of the Swedish Health Institute, Washington, D. C.; late Instructor in Physical Culture and Gymnastics at the Johns Hopkins University, Baltimore, Md.; Instructor of Swedish and German Gymnastics at Harvard University's Summer School, 1891.

This excellent little volume treats this very important subject in a practical manner. Full instructions are given regarding the mode of applying the Swedish Movement and Massage Treatment in various diseases and conditions of the human system with the greatest degree of effectiveness. Professor Nissen is the best authority in the United States upon the practical phase of this subject, and his book is indispensable to every physician who wishes to *know how* to use these valuable handmaids of medicine.

Illustrated with 29 Original Wood-Engravings. In one 12mo volume of 128 Pages. Neatly bound in Cloth.

Price, post-paid, in the United States and Canada, $1.00, net; Great Britain, 6s.; France, 6 fr. 20.

This manual is valuable to the practitioner, as it contains a terse description of a subject but too little understood in this country. . . . The book is got up very creditably.—*N. Y. Med. Jour.*

The present volume is a modest account of the application of the Swedish Movement and Massage Treatment, in which the technique of the various procedures are clearly stated as well as illustrated in a very excellent manner. —*North American Practitioner.*

This attractive little book presents the subject in a very practical shape, and makes it possible for every physician to understand at least how it is applied, if it does not give him dexterity in the art of its application.—*Chicago Med. Times.*

Physicians' Interpreter.

IN FOUR LANGUAGES (ENGLISH, FRENCH, GERMAN, AND ITALIAN). SPECIALLY ARRANGED FOR DIAGNOSIS BY M. VON V.

The object of this little work is to meet a need often keenly felt by the busy physician, namely, the need of some quick and reliable method of communicating intelligibly with patients of those nationalities and languages unfamiliar to the practitioner. The plan of the book is a systematic arrangement of questions upon the various branches of Practical Medicine, and each question is so worded that the only answer required of the patient is merely Yes or No. The questions are all numbered, and a complete Index renders them always available for quick reference. The book is written by one who is well versed in English, French, German, and Italian, being an excellent teacher in all those languages, and who has also had considerable hospital experience. Bound in Full Russia Leather, for carrying in the pocket. Size, 5x2¾ inches. 206 pages.

Price, post-paid, in United States and Canada, $1.00, net; Great Britain, 6s.; France, 6 fr. 20.

Many other books of the same sort, with more extensive vocabularies, have been published, but, from their size, and from their being usually devoted to equivalents in English and one other language only, they have not had the advantage which is pre-eminent in this—convenience. It is handsomely printed, and bound in flexible red leather in the form of a diary. It would scarcely make itself felt in one's hip-pocket, and would insure its bearer against any ordinary conversational difficulty in dealing with foreign-speaking people, who are constantly coming into our city hospitals.—*New York Medical Journal.*

This little volume is one of the most ingenious aids to the physician which we have seen. We heartily commend the book to any one who, being without a knowledge of the foreign languages, is obliged to treat those who do not know our own language.—*St. Louis Courier of Medicine.*

Physician's All-Requisite Time- and Labor-Saving Account-Book.

BEING A LEDGER AND ACCOUNT-BOOK FOR PHYSICIANS' USE, MEETING ALL THE REQUIREMENTS OF THE LAW AND COURTS.

Designed by WILLIAM A. SEIBERT, M.D , of Easton, Pa.

Probably no class of people lose more money through carelessly kept accounts and overlooked or neglected bills than physicians. Often detained at the bedside of the sick until late at night, or deprived of even a modicum of rest, it is with great difficulty that he spares the time or puts himself in condition to give the same care to his own financial interests that a merchant, a lawyer, or even a farmer devotes. It is then plainly apparent that a system of bookkeeping and accounts that, without sacrificing accuracy, but, on the other hand, ensuring it, at the same time relieves the keeping of a physician's book of half their complexity and two-thirds the labor, is a convenience which will be eagerly welcomed by thousands of overworked physicians. Such a system has at last been devised, and we take pleasure in offering it to the profession in the form of The Physician's All-Requisite Time- and Labor- Saving Account-Book.

There is no exaggeration in stating that this Account-Book and Ledger reduces the labor of keeping your accounts more than one half, and at the same time secures the greatest degree of accuracy. We may mention a few of the superior advantages of The Physician's All-Requisite Time- and Labor- Saving Account-Book, as follows :—

First—Will meet all the requirements of the law and courts.

Second—Self-explanatory; no cipher code.

Third—Its completeness without sacrificing anything.

Fourth—No posting; one entry only.

Fifth—Universal; can be commenced at any time of the year, and can be continued indefinitely until every account is filled.

Sixth—Absolutely no waste of space.

Seventh—One person must needs be sick every day of the year to fill his account, or might be ten years about it and require no more than the space for one account in this ledger.

Eighth—Double the number and many times more than the number of accounts in any similar book; the 300-page book contains space for 900 accounts, and the 600-page book contains space for 1800 accounts.

Ninth—There are no smaller spaces.

Tenth—Compact without sacrificing completeness; every account complete on same page—a decided advantage and recommendation.

Eleventh—Uniform size of leaves.

Twelfth—The statement of the most complicated account is at once before you at any time of month or year—in other words, the account itself as it stands is its simplest statement.

Thirteenth—No transferring of accounts, balances, etc.

To all physicians desiring a quick, accurate, and comprehensive method of keeping their accounts, we can safely say that no book as suitable as this one has ever been devised. A descriptive circular showing the plan of the book will be sent on application.

NET PRICES, SHIPPING EXPENSES PREPAID.

	In U.S.	Canada (duty paid).	Great Britain.	France.
No. 1. 300 Pages, for 900 Accounts per Year, Size 10x12, Bound in ¾-Russia, Raised Back-Bands, Cloth Sides,	$5.00	$5.50	28s.	30 fr. 30
No. 2. 600 Pages, for 1800 Accounts per Year, Size 10x12, Bound in ¾-Russia, Raised Back-Bands, Cloth Sides,	8.00	8.80	42s.	49 fr. 40

PRICE and EAGLETON

Three Charts of the Nervo-Vascular System.

PART I.—THE NERVES.   PART II.—THE ARTERIES.
PART III.—THE VEINS.

A New Edition. Revised and Perfected. Arranged by W. HENRY PRICE, M.D., and S. POTTS EAGLETON, M.D. Endorsed by leading anatomists. Clearly and beautifully printed upon extra durable paper.

PART I. The Nerves.—Gives in a clear form not only the Cranial and Spinal Nerves, showing the formation of the different Plexuses and their branches, but also the complete distribution of the SYMPATHETIC NERVES.

PART II. The Arteries.—Gives a unique grouping of the Arterial system, showing the divisions and subdivisions of all the vessels, beginning from the heart and tracing their CONTINUOUS distribution to the periphery, and showing at a glance the terminal branches of each artery.

PART III. The Veins.—Shows how the blood from the periphery of the body is gradually collected by the larger veins, and these coalescing forming still larger vessels, until they finally trace themselves into the Right Auricle of the heart.

It is therefore readily seen that "The Nervo-Vascular System of Charts" offers the following superior advantages :—

1. It is the only arrangement which combines the Three Systems, and yet each is perfect and distinct in itself.

2. It is the only instance of the Cranial, Spinal, and Sympathetic Nervous Systems being represented on one chart.

3. From its neat size and clear type, and being printed only upon one side, it may be tacked up in any convenient place, and is always ready for freshening up the memory and reviewing for examination.

Price, post-paid, in United States and Canada, 50 cents, net, complete ; in Great Britain, 3s. 6d. ; in France, 3 fr. 60.

For the student of anatomy there can possibly be no more concise way of acquiring a knowledge of the nerves, veins, and arteries of the human system. It presents at a glance their trunks and branches in the great divisions of the body. It will save a world of tedious reading, and will impress itself on the mind as no ordinary *vade mecum*, even, could.

Its price is nominal and its value inestimable. No student should be without it.—*Pacific Record of Medicine and Surgery.*

These are three admirably arranged charts for the use of students, to assist in memorizing their anatomical studies.—*Buffalo Med. and Surg. Jour.*

PURDY

Diabetes: Its Cause, Symptoms and Treatment

By CHAS. W. PURDY, M.D. (Queen's University), Honorary Fellow of the Royal College of Physicians and Surgeons of Kingston ; Member of the College of Physicians and Surgeons of Ontario ; Author of "Bright's Disease and Allied Affections of the Kidneys ;" Member of the Association of American Physicians ; Member of the American Medical Association ; Member of the Chicago Academy of Sciences, etc.

CONTENTS.—Section I. Historical, Geographical, and Climatological Considerations of Diabetes Mellitus. II. Physiological and Pathological Considerations of Diabetes Mellitus III. Etiology of Diabetes Mellitus. IV. Morbid Anatomy of Diabetes Mellitus. V. Symptomatology of Diabetes Mellitus. VI. Treatment of Diabetes Mellitus. VII. Clinical Illustrations of Diabetes Mellitus. VIII. Diabetes Insipidus ; Bibliography.

12mo. Dark Blue Extra Cloth. Nearly 200 pages. With Clinical Illustrations. *No. 8 in the Physicians' and Students' Ready-Reference Series.*

Price, post-paid, in the United States and Canada, $1.25, net ; in Great Britain, 6s. 6d. ; in France, 7 fr. 75.

This will prove a most entertaining as well as most interesting treatise upon a disease which frequently falls to the lot of every practitioner. The work has been written with a special view of bringing out the features of the disease as it occurs in the United States. The author has very judiciously arranged the little volume, and it will offer many pleasant attractions to the practitioner.—*Nashville Journal of Medicine and Surgery.*

While many monographs have been pub-

lished which have dealt with the subject of diabetes, we know of none which so thoroughly considers its relations to the geographical conditions which exist in the United States, nor which is more complete in its summary of the symptomatology and treatment of this affection. A number of tables, showing the percentage of sugar in a very large number of alcoholic beverages, adds very considerably to the value of the work.—*Medical News.*

REMONDINO

Circumcision : Its History, Modes of Operation, Etc.

FROM THE EARLIEST TIMES TO THE PRESENT; WITH A HISTORY OF EUNUCHISM, HERMAPHRODISM, ETC., AS OBSERVED AMONG ALL RACES AND NATIONS; ALSO A DESCRIPTION OF THE DIFFERENT OPERATIVE METHODS OF MODERN SURGERY PRACTICED UPON THE PREPUCE.

By P. C. REMONDINO, M.D. (Jefferson) ; Member of the American Medical Association; Member of the American Public Health Association; Vice-President of the State Medical Society of California, and of the Southern California Medical Society, etc., etc.

No. 11 in the Physicians' and Students' Ready-Reference Series. About 350 pages. 12mo. Handsomely bound in Dark-Blue Cloth. JUST READY.

Price, post-paid, in the United States and Canada, $1.25, net ; in Great Britain, 8s. 6d. ; in France, 7 fr. 75. Cheap Edition (paper binding), United States and Canada, 50 cents, net, post-paid; Great Britain, 4s. 3d. ; France, 4 fr. 20.

ROHÉ

Text-Book of Hygiene.

A COMPREHENSIVE TREATISE ON THE PRINCIPLES AND PRACTICE OF PRE-VENTIVE MEDICINE FROM AN AMERICAN STAND-POINT.

By GEORGE H. ROHÉ, M.D., Professor of Obstetrics and Hygiene in the College of Physicians and Surgeons, Baltimore ; Member of the American Public Health Association, etc.

Every Sanitarian should have Rohé's "Text-Book of Hygiene" as a work of reference. Of this New (second) edition, one of the best qualified judges, namely, Albert L. Gihon, M.D., Medical Director, U. S. Navy, in charge of U. S. Naval Hospital, Brooklyn, N. Y., and ex-President of the American Public Health Association, writes : " It is the most admirable, concise *résumé* of the facts of Hygiene with which I am acquainted. Prof. Rohé's attractive style makes the book so readable that no better presentation of the important place of Pre-ventive Medicine, among their studies, can be desired for the younger members, especially, of our profession.

Second Edition, thoroughly revised and largely rewritten, with many illustrations and valuable tables. In one handsome Royal Octavo volume of over 400 pages, bound in Extra Cloth.

Price, post-paid, in United States, $2.50, net ; Canada (duty paid), $2.75, net ; Great Britain, 14s. ; France, 16 fr. 20.

In short, the work contains brief and prac-tical articles on hygienic regulation of life, under almost all conditions One prominent feature is that there are no superfluous words; every sentence is direct to the point sought. It is, therefore, easy reading, and conveys very much information in little space.—*The Pacific Record of Medicine and Surgery.*

Truly a most excellent and valuable work, comprising the accepted facts in regard to preventive medicine, clearly stated and well arranged. It is unquestionably a work that should be in the hands of every physician in the country, and medical students will find it a most excellent and valuable text-book.—*The Southern Practitioner.*

The first edition was rapidly exhausted, and the book justly became an authority in regard to physicians and sanitary officers, and a text-book very generally adopted in the colleges throughout America. The second edition is a great improve-ment over the first, all of the matter being thor-

oughly revised, much of it being rewritten, and many additions being made. The size of the book is increased one hundred pages. The book has the original recommendation of being a handsomely-bound, clearly-printed octavo volume, profusely illustrated with re-liable references for every branch of the subject matter.—*Medical Record.*

The wonder is how Prof. Rohé has made the book so readable and entertaining with so much matter necessarily condensed. The book is well printed with good, clear type, is attractive in appearance, and contains a number of valuable tables and illustrations that must be of decided aid to the student, if not to the general practitioner and health officer. Altogether, the manual is a good ex-ponent of hygiene and sanitary science from the present American stand-point, and will repay with pleasure and profit any time that may be given to its perusal.—*University Medi-cal Magazine.*

SAJOUS

HAY FEVER And Its Successful Treatment by Superficial Organic Alteration of the Nasal Mucous Membrane.

By CHARLES E. SAJOUS, M.D., formerly Lecturer on Rhinology and Laryngology in Jefferson Medical College; Vice-President of the American Laryngological Association; Officer of the Academy of France and of Public Instruction of Venezuela; Corresponding Member of the Royal Society of Belgium, of the Medical Society of Warsaw (Poland), and of the Society of Hygiene of France; Member of the American Philosophical Society, etc., etc.

With 13 Engravings on Wood. 103 pages. 12mo. Bound in Cloth. Beveled Edges.

Price, post-paid, in the United States and Canada, $1.00, net; in Great Britain, 6s.; France, 6 fr. 20.

SANNÉ

Diphtheria, Croup: Tracheotomy and Intubation.

FROM THE FRENCH OF A. SANNÉ.

Translated and enlarged by HENRY Z. GILL, M.D., LL.D., late Professor of Surgery in Cleveland, Ohio.

SANNÉ'S work is quoted, directly or indirectly, by every writer since its publication, as the highest authority, statistically, theoretically, and practically. The translator, having given special study to the subject for many years, has added over fifty pages, including the Surgical Anatomy, Intubation, and the recent progress in other branches, making it, beyond question, the most complete work extant on the subject of Diphtheria in the English language.

Facing the title-page is found a very fine Colored Lithograph Plate of the parts concerned in Tracheotomy. Next follows an illustration of a cast of the entire Trachea, and bronchi to the third or fourth division, in one piece, taken from a photograph of a case in which the cast was expelled during life from a patient sixteen years old. This is the most complete cast of any one recorded.

Over fifty other illustrations of the surgical anatomy of instruments, etc., add to the practical value of the work.

A full Index accompanies the enlarged volume, also a List of Authors; making altogether a very handsome illustrated volume of over 680 pages.

	United States.	Canada (duty paid).	Great Britain.	France.
Price, post-paid, Cloth, -	$4.00, Net	$4.40, Net	22s. 6d.	24 fr. 60
" " Leather,	5.00, "	5.50, "	28s.	30 fr. 30

The subject of intubation, so recently revived in this country, receives a very careful and impartial discussion at the hands of the translator, and a most valuable chapter on the prophylaxis of diphtheria and croup closes the volume. Altogether the book is one that is valuable and timely, and one that should be in the hands of every general prac-titioner.—*St. Louis Med. and Surgical Journ.*

Diphtheria having become such a prevalent, wide-spread, and fatal disease, no general practitioner can afford to be without this work. It will aid in preventive measures, stimulate promptness in application of, and efficiency in, treatment.—*Southern Practitioner.*

SENN

Principles of Surgery.

By N. SENN, M.D., PH.D., Professor of Practice of Surgery and Clinical Surgery in Rush Medical College, Chicago, Ill.; Professor of Surgery in the Chicago Polyclinic; Attending Surgeon to the Milwaukee Hospital; Consulting Surgeon to the Milwaukee County Hospital and to the Milwaukee County Insane Asylum.

This work, by one of America's greatest surgeons, is thoroughly COMPLETE; its clearness and brevity of statement are among its conspicuous merits. The author's long, able, and conscientious researches in every direction in this important field are a guarantee, of unusual trustworthiness, that every branch of the subject is treated authoritatively and in such a manner as to bring the greatest gain in knowledge to the Practitioner and Student.

In one handsome Royal Octavo volume, with 109 fine Wood-Engravings and 624 pages.

	United States.	Canada (duty paid)	Great Britain.	France.
Price in Cloth,	$4.50, Net	$5.00, Net	24s. 6d.	27 fr. 20
" Sheep or ½-Russia,	5.50 "	6.10 "	30s.	33 fr. 10

OPINIONS AND CRITICISMS:

STEPHEN SMITH, M.D., Professor of Clinical Surgery Medical Department University of the City of New York, writes:—"I have examined the work with great satisfaction, and regard it as a most valuable addition to American Surgical literature. There has long been great need of a work on the principles of surgery which would fully illustrate the present advanced state of knowledge of the various subjects embraced in this volume. The work seems to me to meet this want admirably."

LEWIS A. SAYRE, M.D., Professor Orthopedic Surgery Bellevue Hospital Medical College, New York, writes:—"My Dear Doctor Senn: Your very valuable work on surgery, sent to me some time since, I have studied with great satisfaction and improvement. I congratulate you most heartily on having produced the most classical and practical work on surgery yet published."

FRANK J. LUTZ, M.D., St. Louis, Mo., says:—"It seems incredible that those who pretend to teach have done without such a guide before, and I do not understand how our students succeeded in mastering the principles of modern surgery by attempting to read our obsolete text-books. American surgery should feel proud of the production, and the present generation of surgeons owe you a debt of gratitude."

W. W. DAWSON, M.D., Cincinnati, Ohio, writes:—"It is a work of great merit, and one greatly needed. Reliable Surgery must be founded upon correct principles."

WM. OSLER, M.D., The Johns Hopkins Hospital, Baltimore, says:—"You certainly have covered the ground thoroughly and well, and with a thoroughness I do not know of in any similar work. I should think it would prove a great boon to the students and also to very many teachers."

J. C. WARREN, M.D., Boston, Mass., writes:—"The book comes at an opportune moment; the old text-books on Surgical Pathology are out of date, and you are filling practically a new field."

The work is systematic and compact, without a fact omitted or a sentence too much, and it not only makes instructive but fascinating reading. A conspicuous merit of Senn's work is his method, his persistent and tireless search through original investigations for additions to knowledge, and the practical character of his discoveries. This combination of the discoverer and the practical man gives a special value to all his work, and is one of the secrets of his fame. No physician in any line of practice can afford to be without Senn's "Principles of Surgery."—*The Review of Insanity and Nervous Diseases.*

Every chapter is a mine of information containing all the recent advances on the subjects presented in such a systematic, instructive and entertaining style that the reader will not willingly lay it aside, but will read and re-read with pleasure and profit.—*Kansas Medical Journal.*

It is a most admirable work in all respects, and should be in the hands of every senior student, general practitioner, and special surgeon.—*Canadian Practitioner.*

After perusing this work on several different occasions we have come to the conclusion that it is a remarkable work by a man of unusual ability. We have never seen anything like it before. The author seems to have had a very large personal experience, which is freely made use of in the text, besides which he is familiar with almost all that has been written in English and German on the above topics. We congratulate Dr. Senn upon the manner in which he has accomplished his task.—*The Canada Medical Record.*

The work is exceedingly practical, as the chapters on the treatment of the various conditions considered are based on sound deductions, are complete, and easily carried out by any painstaking surgeon. Asepsis and antisepsis are exhaustively treated. All in all, the book is a most excellent one, and deserves a place in every well-selected library.—*Medical Record.*

It will prove exceedingly valuable in the diffusion of more thorough knowledge of the subject-matter among English-speaking surgeons. As in the case of all his work, he has done this in a truly admirable manner. Nowhere is there room to criticise the accuracy of Senn's statements, and everywhere is there evidence of a thorough study of the best work of the most eminent men. The book throughout is worthy of the highest praise. It should be adopted as a text-book in all of our schools.—*University Medical Magazine.*

The principles of surgery, as expounded by Dr. Senn, are such as to place the student in the independent position of evolving from them methods of treatment: the master of the principles readily becomes equally a master of practice. And this, of course, is really the whole purpose of the volume.—*Weekly Medical Review.*

SHOEMAKER
Materia Medica and Therapeutics.

WITH ESPECIAL REFERENCE TO THE CLINICAL APPLICATION OF DRUGS. BEING THE SECOND AND LAST VOLUME OF A TREATISE ON MATERIA MEDICA, PHARMACOLOGY, AND THERAPEUTICS, AND AN INDEPENDENT VOLUME UPON DRUGS.

By JOHN V. SHOEMAKER, A.M., M.D., Professor of Materia Medica, Pharmacology, Therapeutics, and Clinical Medicine, and Clinical Professor of Diseases of the Skin in the Medico-Chirurgical College of Philadelphia; Physician to the Medico-Chirurgical Hospital, etc., etc.

This, the second volume of Shoemaker's "Materia Medica, Pharmacology, and Therapeutics," is wholly taken up with the consideration of drugs, each remedy being studied from three points of view, viz.: the Preparations, or Materia Medica; the Physiology and Toxicology, or Pharmacology; and, lastly, its Therapy. It is thoroughly abreast of the progress of Therapeutic Science, and is really an indispensable book to every student and practitioner of medicine.

Royal Octavo, about 675 pages. Thoroughly and carefully indexed.

Price, in United States, post-paid, Cloth, $3.50; Sheep, $4.50, net. Canada (duty paid), Cloth, $4.00; Sheep, $5.00, net. Great Britain, Cloth, 20s.; Sheep, 26s. France, Cloth, 22 fr. 40; Sheep, 28 fr. 60.

The first volume of this work is devoted to Pharmacy, General Pharmacology, and Therapeutics, and remedial agents not properly classed with drugs. Royal Octavo, 354 pages.

Price of Volume I, post-paid, in United States, Cloth, $2.50, net; Sheep, $3.25, net. Canada, duty paid, Cloth, $2.75, net; Sheep, $3.60, net. Great Britain, Cloth, 14s.; Sheep, 18s. France, Cloth, 16 fr. 20; Sheep, 20 fr. 20. The volumes are sold separately.

SHOEMAKER'S TREATISE ON MATERIA MEDICA, PHARMACOLOGY, AND THERA-PEUTICS STANDS ALONE.

(1) Among Materia Medica text-books, in that it includes every officinal drug and every preparation contained in the United States Pharmacopœia.

(2) In that it is the only work on therapeutics giving the strength, composition, and dosage of every officinal preparation.

(3) In giving the latest investigations with regard to the physiological action of drugs and the most recent applications in therapeutics.

(4) In combining with officinal drugs the most reliable reports of the actions and uses of all the noteworthy new remedies, such as acetanilid, antipyrin, bromoform, exalgin, pyoktannin, pyridin, somnal, sperminæ (Brown-Séquard), tuberculin (Koch's lymph), sulphonal, thiol, urethan, etc., etc.

(5) As a complete encyclopædia of modern therapeutics in condensed form, arranged alphabetically for convenience of reference for either physician, dentist, or pharmacist, when immediate information is wanted concerning the action, composition, dose, or antidotes for any officinal preparation or new remedy.

(6) In giving the physical characters and chemical formulæ of the new remedies, especially the recently-introduced antipyretics and analgesics.

(7) In the fact that it gives special attention to the consideration of the diagnosis and treatment of poisoning by the more active drugs, both officinal and non-officinal.

(8) And unrivaled in the number and variety of the prescriptions and practical formulæ, representing the latest achievements of clinical medicine.

(9) In that, while summarizing foreign therapeutical literature, it fully recognizes the work done in this department by American physicians. It is an epitome of the present state of American medical practice, which is universally acknowledged to be the best practice.

(10) Because it is the most complete, convenient, and compendious work of reference, being, in fact, a companion to the United States Pharmacopœia, a drug-encyclopædia, and a therapeutic hand-book all in one volume.

The value of the book lies in the fact that it contains all that is authentic and trustworthy about the host of new remedies which have deluged us in the last five years. The pages are remarkably free from useless information. The author has done well in following the alphabetical order.—*N. Y. Med. Record.*

In perusing the pages devoted to the special consideration of drugs, their pharmacology, physiological action, toxic action, and therapy, one is constantly surprised at the amount of material compressed in so limited a space. The book will prove a valuable addition to the physician's library.—*Occidental Med. Times.*

It is a meritorious work, with many unique features. It is richly illustrated by well-tried prescriptions showing the practical application of the various drugs discussed. In short, this work makes a pretty complete encyclopædia of the science of therapeutics, conveniently arranged for handy reference.—*Med. World.*

SMITH

Physiology of the Domestic Animals.

A TEXT-BOOK FOR VETERINARY AND MEDICAL STUDENTS AND PRACTITIONERS.

By ROBERT MEADE SMITH, A.M., M.D., Professor of Comparative Physiology in University of Pennsylvania; Fellow of the College of Physicians and Academy of the Natural Sciences, Philadelphia; of American Physiological Society; of the American Society of Naturalists, etc.

This new and important work, the most thoroughly complete in the English language on this subject, treats of the physiology of the domestic animals in a most comprehensive manner, especial prominence being given to the subject of foods and fodders, and the character of the diet for the herbivora under different conditions, with a full consideration of their digestive peculiarities. Without being overburdened with details, it forms a complete text-book of physiology adapted to the use of students and practitioners of both veterinary and human medicine. This work has already been adopted as the Text-Book on Physiology in the Veterinary Colleges of the United States, Great Britain, and Canada. In one Handsome Royal Octavo Volume of over 950 pages, profusely illustrated with more than 400 Fine Wood-Engravings and many Colored Plates.

	United States.	Canada (duty paid)	Great Britain.	France
Price, Cloth, - -	$5.00, Net	$5.50, Net	28s.	30 fr. 30
" Sheep, - -	6.00 "	6.60 "	32s.	36 fr. 20

A. LIAUTARD, M.D., H.F.R.C., V.S., Professor of Anatomy, Operative Surgery, and Sanitary Medicine in the American Veterinary College, New York, writes:—"I have examined the work of Dr. R. M. Smith on the 'Physiology of the Domestic Animals,' and consider it one of the best additions to veterinary literature that we have had for some time."

E. M. READING, A.M., M.D., Professor of Physiology in the Chicago Veterinary College, writes:—"I have carefully examined the 'Smith's Physiology,' published by you, and like it. It is comprehensive, exhaustive, and complete, and is especially adapted to those who desire to obtain a full knowledge of the principles of physiology, and are not satisfied with a mere smattering of the cardinal points."

Dr. Smith's presentment of his subject is as brief as the status of the science permits, and to this much-desired conciseness he has added an equally welcome clearness of statement. The illustrations in the work are exceedingly good, and must prove a valuable aid to the

full understanding of the text.—*Journal of Comparative Medicine and Surgery.*

Veterinary practitioners and graduates will read it with pleasure. Veterinary students will readily acquire needed knowledge from its pages, and veterinary schools, which would be well equipped for the work they aim to perform, cannot ignore it as their text-book in physiology.—*American Veterinary Review.*

Altogether, Professor Smith's "Physiology of the Domestic Animals" is a happy production, and will be hailed with delight in both the human medical and veterinary medical worlds. It should find its place, besides, in all agricultural libraries.—PAUL PAQUIN, M.D., V.S., in the *Weekly Medical Review.*

The author has judiciously made the nutritive functions the strong point of the work, and has devoted special attention to the subject of foods and digestion. In looking through other sections of the work, it appears to us that a just proportion of space is assigned to each, in view of their relative importance to the practitioner.—*London Lancet.*

SOZINSKEY

Medical Symbolism. Historical Studies in the Arts of Healing and Hygiene.

By THOMAS S. SOZINSKEY, M.D., PH.D., Author of "The Culture of Beauty," "The Care and Culture of Children," etc.

12mo. Nearly 200 pages. Neatly bound in Dark-Blue Cloth. Appropriately illustrated with upward of thirty (30) new Wood-Engravings. *No. 9 in the Physicians' and Students' Ready-Reference Series.*

Price, post-paid, in United States and Canada, $1.00, net; Great Britain, 6s.; France, 6 fr. 20.

He who has not time to more fully study the more extended records of the past, will highly prize this little book. Its interesting discourse upon the past is full of suggestive thought.—*American Lancet.*

Like an oasis in a dry and dusty desert of medical literature, through which we wearily stagger, is this work devoted to medical symbolism and mythology. As the author aptly quotes: "What some light braines may esteem as foolish toyes, deeper judgments can and

will value as sound and serious matter."—*Canadian Practitioner.*

In the volume before us we have an admirable and successful attempt to set forth in order those medical symbols which have come down to us, and to explain on historical grounds their significance. An astonishing amount of information is contained within the covers of the book, and every page of the work bears token of the painstaking genius and erudite mind of the now unhappily deceased author.—*London Lancet.*

STEWART

Obstetric Synopsis.

By John S. Stewart, M.D., formerly Demonstrator of Obstetrics and Chief Assistant in the Gynæcological Clinic of the Medico-Chirurgical College of Philadelphia; with an introductory note by William S. Stewart, A.M., M.D., Professor of Obstetrics and Gynæcology in the Medico-Chirurgical College of Philadelphia.

By students this work will be found particularly useful. It is based upon the teachings of such well-known authors as Playfair, Parvin, Lusk, Galabin, and Cazeaux and Tarnier, and contains much new and important matter of great value to both student and practitioner.

With 42 Illustrations. 202 pages. 12mo. Handsomely bound in Dark-Blue Cloth. *No. 1 in the Physicians' and Students' Ready-Reference Series.*

Price, post-paid, in the United States and Canada, $1.00, net; in Great Britain, 6s.; France, 6 fr. 20.

Delaskie Miller, M.D., Professor of Obstetrics, Rush Medical College, Chicago, Ill., says:—"I have examined the 'Obstetric Synopsis,' by John S. Stewart, M.D., and it gives me pleasure to characterize the work as systematic, concise, perspicuous, and authentic. Among manuals it is one of the best."

It is well written, excellently illustrated, and fully up to date in every respect. Here we find all the essentials of Obstetrics in a nutshell, Anatomy, Embryology, Physiology, Pregnancy, Labor, Puerperal State, and Obstetric Operations all being carefully and ac-

curately described.—*Buffalo Medical and Surgical Journal.*

It is clear and concise. The chapter on the development of the ovum is especially satisfactory. The judicious use of bold-faced type for headings and italics for important statements gives the book a pleasing typographical appearance.—*Medical Record.*

This volume is done with a masterly hand. The scheme is an excellent one. The whole is freely and most admirably illustrated with well-drawn, new engravings, and the book is of a very convenient size.—*St. Louis Medical and Surgical Journal.*

ULTZMANN

The Neuroses of the Genito-Urinary System in the Male.

With Sterility and Impotence.

By Dr. R. Ultzmann, Professor of Genito-Urinary Diseases in the University of Vienna. Translated, with the author's permission, by Gardner W. Allen, M.D., Surgeon in the Genito-Urinary Department, Boston Dispensary.

Full and complete, yet terse and concise, it handles the subject with such a vigor of touch, such a clearness of detail and description, and such a directness to the result, that no medical man who once takes it up will be content to lay it down until its perusal is complete,—nor will one reading be enough.

Professor Ultzmann has approached the subject from a somewhat different point of view from most surgeons, and this gives a peculiar value to the work. It is believed, moreover, that there is no convenient hand-book in English treating in a broad manner the Genito-Urinary Neuroses.

Synopsis of Contents.—First Part—I. Chemical Changes in the Urine in Cases of Neuroses. II. Neuroses of the Urinary and of the Sexual Organs, classified as: (1) Sensory Neuroses; (2) Motor Neuroses; (3) Secretory Neuroses. Second Part—Sterility and Impotence. The treatment in all cases is described clearly and minutely.

Illustrated. 12mo. Handsomely bound in Dark-Blue Cloth. *No. 4 in the Physicians' and Students' Ready-Reference Series.*

Price, post-paid, in the United States and Canada, $1.00, net; in Great Britain, 6s.; in France, 6 fr. 20.

This book is to be highly recommended, owing to its clearness and brevity. Altogether, we do not know of any book of the same size which contains so much useful information in such a short space.—*Medical News.*

Its scope is large, not being confined to the one condition,—neurasthenia,—but embracing all of the neuroses, motor and sensory, of the genito-urinary organs in the male. No one who has read after Dr. Ultzmann need be re-

minded of his delightful manner of presenting his thoughts, which ever sparkle with originality and appositeness.—*Weekly Med. Review.*

It engenders sound pathological teaching, and will aid in no small degree in throwing light on the management of many of the difficult and more refractory cases of the classes to which these essays especially refer.—*The Medical Age.*

WHEELER

Abstracts of Pharmacology.

By H. A. WHEELER, M.D. (Registered Pharmacist, No. 3468, Iowa). Prepared for the use of Physicians and Pharmacists, and especially for the use of Students of Medicine and Pharmacy, who are preparing for Examination in Colleges and before State Boards of Examiners.

This book does not contain questions and answers, but solid pages of abstract information. It will be an almost indispensable companion to the practicing Pharmacist, and a very useful reference-book to the Physician. It contains a brief but thorough explanation of all terms and processes used in practical pharmacy, an abstract of all that is essential to be known of each officinal drug, its preparations and therapetic action, with doses; in Chemistry and Botany, much that is useful to the Physician and Pharmacist; a general working formula for each class and an abstract formula for each officinal preparation, and many of the more popular unofficinal ones, together with their doses; also many symbolic formulas; a list of abbreviations used in prescription writing; rules governing incompatibilities; a list of Solvents; tests for the more common drugs; the habitat and best time for gathering plants to secure their medical properties.

The book contains 180 pages, 5¼ x 8 inches, closely printed and on the best paper, nicely and durably bound, containing a greater amount of information on the above topics than any other work for the money.

Price, post-paid, in United States and Canada, $1.50, net; in Great Britain, 8s. 6d.; in France, 9 fr. 35.

YOUNG

Synopsis of Human Anatomy.

BEING A COMPLETE COMPEND OF ANATOMY, INCLUDING THE ANATOMY OF THE VISCERA, AND NUMEROUS TABLES.

By JAMES K. YOUNG, M.D., Instructor in Orthopædic Surgery and Assistant Demonstrator of Surgery, University of Pennsylvania; Attending Orthopædic Surgeon, Out-Patient Department, University Hospital, etc.

While the author has prepared this work especially for students, sufficient descriptive matter has been added to render it extremely valuable to the busy practitioner, particularly the sections on the Viscera, Special Senses, and Surgical Anatomy.

The work includes a complete account of Osteology, Articulations and Ligaments, Muscles, Fascias, Vascular and Nervous Systems, Alimentary, Vocal, and Respiratory and Genito-Urinary Apparatus, the Organs of Special Sense, and Surgical Anatomy.

In addition to a most carefully and accurately prepared text, wherever possible, the value of the work has been enhanced by tables to facilitate and minimize the labor of students in acquiring a thorough knowledge of this important subject. The section on the teeth has also been especially prepared to meet the requirements of students of dentistry.

Illustrated with 76 Wood-Engravings. 390 pages. 12mo. *No. 3 in the Physicians' and Students' Ready-Reference Series.*

Price, post-paid, in United States and Canada $1.40, net; Great Britain, 8s. 6d.; France, 9 fr. 25.

Every unnecessary word has been excluded, out of regard to the very limited time at the medical student's disposal. It is also good as a reference-book, as if presents the facts about which he wishes to refresh his memory in the briefest manner consistent with clearness.—*New York Medical Journal.*

As a companion to the dissecting-table, and a convenient reference for the practitioner, it has a definite field of usefulness.—*Pittsburgh Medical Review.*

The book is much more satisfactory than the " remembrances " in vogue, and yet is not too cumbersome to be carried around and read at odd moments—a property which the student will readily appreciate. — *Weekly Medical Review.*

WITHERSTINE

The International Pocket Medical Formulary

ARRANGED THERAPEUTICALLY.

By C. SUMNER WITHERSTINE, M.S., M.D., Associate Editor of the "Annual of the Universal Medical Sciences;" Visiting Physician of the Home for the Aged, Germantown, Philadelphia; Late House-Surgeon Charity Hospital, New York.

More than 1800 formulæ from several hundred well-known authorities. With an Appendix containing a Posological Table, the newer remedies included; Important Incompatibles; Tables on Dentition and the Pulse; Table of Drops in a Fluidrachm and Doses of Laudanum graduated for age; Formulæ and Doses of Hypodermatic Medication, including the newer remedies; Uses of the Hypodermatic Syringe; Formulæ and Doses for Inhalations, Nasal Douches, Gargles, and Eye-washes; Formulæ for Suppositories; Use of the Thermometer in Disease; Poisons, Antidotes, and Treatment; Directions for Post-Mortem and Medico-Legal Examinations; Treatment of Asphyxia, Sun-stroke, etc.; Antiemetic Remedies and Disinfectants; Obstetrical Table; Directions for Ligations of Arteries; Urinary Analysis; Table of Eruptive Fevers; Motor Points for Electrical Treatment, etc.

This work, the best and most complete of its kind, contains about 275 printed pages, besides extra blank leaves—the book being interleaved throughout—elegantly printed, with red lines, edges, and borders; with illustrations. Bound in leather, with side flap.

It is a handy book of reference, replete with the choicest formulæ (over 1800 in number) of more than six hundred of the most prominent classical writers and modern practitioners.

The remedies given are not only those whose efficiency has stood the test of time, but also the newest and latest discoveries in pharmacy and medical science, as prescribed and used by the best-known American and foreign modern authorities.

It contains the latest, largest (66 formulæ), and most complete collection of hypodermatic formulæ (including the latest new remedies) ever published, with doses and directions for their use in over fifty different diseases and diseased conditions.

Its appendix is brimful of information, invaluable in office work, emergency cases, and the daily routine of practice.

It is a reliable friend to consult when, in a perplexing or obstinate case, the usual line of treatment is of no avail. (A hint or a help from the best authorities, as to choice of remedies, correct dosage, and the eligible, elegant, and most palatable mode of exhibition of the same.)

It is compact, elegantly printed and bound, well illustrated, and of convenient size and shape for the pocket.

The alphabetical arrangement of the diseases and a thumb-letter index render reference rapid and easy.

Blank leaves, judiciously distributed throughout the book, afford a place to record and index favorite formulæ.

As a *student*, the physician needs it for study, collateral reading, and for recording the favorite prescriptions of his professors, in lecture and clinic; as a *recent graduate*, he needs it as a reference hand-book for daily use in prescribing (gargles, nasal douches, inhalations, eye-washes, suppositories, incompatibles, poisons, etc.); as an *old practitioner*, he needs it to refresh his memory on old remedies and combinations, and for information concerning newer remedies and more modern approved plans of treatment.

No live, progressive medical man can afford to be without it.

Annual of the Universal Medical Sciences.

A YEARLY REPORT OF THE PROGRESS OF THE GENERAL SANITARY
SCIENCES THROUGHOUT THE WORLD.

Edited by CHARLES E. SAJOUS, M.D., formerly Lecturer on Laryngology
and Rhinology in Jefferson Medical College, Philadelphia, etc., and Seventy
Associate Editors, assisted by over Two Hundred Corresponding Editors and
Collaborators. In Five Royal Octavo Volumes of about 500 pages each, bound
in Cloth and Half-Russia, Magnificently Illustrated with Chromo-Lithographs,
Engravings, Maps, Charts, and Diagrams. Being intended to enable any physician to possess, at a moderate cost, a complete Contemporary History of Universal
Medicine, edited by many of America's ablest teachers, and superior in every
detail of print, paper, binding, etc., a befitting continuation of such great works
as "Pepper's System of Medicine," "Ashhurst's International Encyclopædia of
Surgery," "Buck's Reference Hand-Book of the Medical Sciences."

**SOLD ONLY BY SUBSCRIPTION, OR SENT DIRECT ON RECEIPT OF PRICE,
SHIPPING EXPENSES PREPAID.**

Subscription Price per Year (including the "SATELLITE" for one year):
In United States, Cloth, 5 vols., Royal Octavo, $15.00; Half-Russia, 5 vols.,
Royal Octavo, $20.00. Canada (duty paid), Cloth, $16.50; Half-Russia,
$22.00. Great Britain, Cloth, £4 7s.; Half-Russia, £5 15s. France, Cloth,
93 fr. 95; Half-Russia, 124 fr. 35.

THE SATELLITE of the "Annual of the Universal Medical Sciences." A
Monthly Review of the most important articles upon the practical branches of
Medicine appearing in the medical press at large, edited by the Chief Editor of
the ANNUAL and an able staff. Published in connection with the ANNUAL, and
for its Subscribers Only.

Editorial Staff of the Annual of the Universal Medical Sciences.

CONTRIBUTORS TO SERIES 1888, 1889, 1890, 1891.

EDITOR-IN-CHIEF, CHARLES E. SAJOUS, M.D., PHILADELPHIA.

SENIOR ASSOCIATE EDITORS.

Agnew, D. Hayes, M.D., LL.D., Philadelphia, series of 1888, 1889.

Baldy, J. M., M.D., Philadelphia, 1891.

Barton, J. M., A.M., M.D., Philadelphia, 1889, 1890, 1891.

Birdsall, W. R., M.D., New York, 1889, 1890, 1891.

Brown, F. W., M.D., Detroit, 1890, 1891.

Bruen, Edward T., M.D., Philadelphia, 1889.

Brush, Edward N., M.D., Philadelphia, 1889, 1890, 1891.

Cohen, J. Solis, M.D., Philadelphia, 1888, 1889, 1890, 1891.

Conner, P. S., M.D., LL.D., Cincinnati, 1888, 1889, 1890, 1891.

Currier, A. F., A.B., M.D., New York, 1889, 1890, 1891.

Davidson, C. C., M.D., Philadelphia, 1888.

Davis, N. S., A.M., M.D., LL.D., Chicago, 1888, 1889, 1890, 1891.

Delafield, Francis, M.D., New York, 1888.

Delavan, D. Bryson, M.D., New York, 1888, 1889, 1890, 1891.

Draper, F. Winthrop, A.M., M.D., New York, 1888, 1889, 1890, 1891.

Dudley, Edward C., M.D., Chicago, 1888.

Ernst, Harold C., A.M., M.D., Boston, 1889, 1890, 1891.

Forbes, William S., M.D., Philadelphia, 1888, 1889, 1891.

Garretson, J. E., M.D., Philadelphia, 1888, 1889.

Gaston, J. McFadden, M.D., Atlanta, 1890, 1891.

Gihon, Albert L., A.M., M.D., Brooklyn, 1888, 1889, 1890, 1891.

Goodell, William, M.D., Philadelphia, 1888, 1889, 1890.

Gray, Landon Carter, M.D., New York, 1890, 1891.

Griffith, J. P. Crozer, M.D., Philadelphia, 1889, 1890, 1891.

Guilford, S. H., D.D.S., Ph.D., Philadelphia, 1888.

Guiteras, John, M.D., Ph.D., Charleston, 1888, 1889.

Hamilton, John B., M.D., LL.D., Washington, 1888, 1889, 1890, 1891.

Hare, Hobart Amory, M.D., B.Sc., Philadelphia, 1888, 1889, 1890, 1891.

Henry, Frederick P., M.D., Philadelphia, 1889, 1890, 1891.

Holland, J. W., M.D., Philadelphia, 1888, 1889, 1890.

Holt, L. Emmett, M.D., New York, 1889, 1890, 1891.

Howell, W. H., Ph.D., M.D., Ann Arbor, 1889, 1890, 1891.

Hun, Henry, M.D., Albany, 1889, 1890.

Hooper, Franklin H., M.D., Boston, 1890, 1891.

Ingals, E. Fletcher, A.M., M.D., Chicago, 1889, 1890, 1891.

Jaggard, W. W., A.M., M.D., Chicago, 1890.

Johnston, Christopher, M.D., Baltimore, 1888, 1889.

Johnston, W. W., M.D., Washington, 1888, 1889, 1890, 1891.

SENIOR ASSOCIATE EDITORS
(CONTINUED).

Keating, John M., M.D., Philadelphia, 1889.
Kelsey, Charles B., M.D., New York, 1888, 1889, 1890, 1891.
Keyes, Edward L., A.M., M.D., New York, 1888, 1889, 1890, 1891.
Knapp, Philip Coombs, M.D., Boston, 1891.
Laplace, Ernest, A.M., M.D., Philadelphia, 1890, 1891.
Lee, John G., M.D., Philadelphia, 1888.
Leidy, Joseph, M.D., LL.D., Philadelphia, 1888, 1889, 1890, 1891.
Longstreth, Morris, M.D., Philadelphia, 1888, 1889, 1890.
Loomis, Alfred L., M.D., LL.D., New York, 1888, 1889.
Lyman, Henry M., A.M., M.D., Chicago, 1888.
McGuire, Hunter, M.D., LL.D., Richmond, 1888.
Manton, Walter P., M.D., F.R.M.S., Detroit, 1888, 1889, 1890, 1891.
Martin, H. Newell, M.D., M.A., Dr.Sc., F.R.S., Baltimore, 1888, 1889.
Matas, Rudolph, M.D., New Orleans, 1890, 1891.
Mears, J. Ewing, M.D., Philadelphia, 1888, 1889, 1890, 1891.
Mills, Charles K., M.D., Philadelphia, 1888.
Minot, Chas. Sedgwick, M.D., Boston, 1888, 1889, 1890, 1891.
Montgomery, E. E., M.D., Philadelphia, 1891.
Morton, Thos. G., M.D., Philadelphia, 1888, 1889.
Mundé, Paul F., M.D., New York, 1888, 1889, 1890, 1891.
Oliver, Charles A., A.M., M.D., Philadelphia, 1889, 1890, 1891.
Packard, John H., A.M., M.D., Philadelphia, 1888, 1889, 1890, 1891.
Parish, Wm. H., M.D., Philadelphia, 1888, 1889, 1890.
Parvin, Theophilus, M.D., LL.D., Philadelphia, 1888, 1889.
Pierce, C. N., D.D.S., Philadelphia, 1888.
Pepper, William, M.D., LL.D., Philadelphia, 1888.
Ranney, Ambrose L., M.D., New York, 1888, 1889, 1890.
Richardson, W. L., M.D., Boston, 1888, 1889.
Rockwell, A. D., A.M., M.D., New York, 1891.
Rohé, Geo. H., M.D., Baltimore, 1888, 1889, 1890, 1891.
Sajous, Chas. E., M.D., Philadelphia, 1888, 1889, 1890, 1891.
Sayre, Lewis A., M.D., New York, 1890, 1891.
Seguin, E. C., M.D., Providence, 1888, 1889, 1890, 1891.
Senn, Nicholas, M.D., Ph.D., Milwaukee, 1888, 1889.
Shakspeare, E. O., M.D., Philadelphia, 1888.
Shattuck, F. C., M.D., Boston, 1890.
Smith, Allen J., A.M., M.D., Philadelphia, 1890, 1891.
Smith, J. Lewis, M.D., New York, 1888, 1889, 1890, 1891.
Spitzka, E. C., M.D., New York, 1888.
Starr, Louis, M.D., Philadelphia, 1888, 1889, 1890, 1891.
Stimson, Lewis A., M.D., New York, 1888, 1889, 1890, 1891.
Sturgis, F. R., M.D., New York, 1888.
Suddath, F. X., A.M., M.D., F.R.M.S., Minneapolis, 1888, 1889, 1890, 1891.
Thomson, William, M.D., Philadelphia, 1888.
Thomson, Wm. H., M.D., New York, 1888.
Tiffany, L. McLane, A.M., M.D., Baltimore, 1890, 1891.
Turnbull, Chas. S., M.D., Ph.D., Philadelphia, 1888, 1889, 1890, 1891.
Tyson, James, M.D., Philadelphia, 1888, 1889, 1890.
Van Harlingen, Arthur, M.D., Philadelphia, 1888, 1889, 1890, 1891.
Vander Veer, Albert, M.D., Ph.D., Albany, 1891.
Whittaker, Jas. T., M.D., Cincinnati, 1888, 1889, 1890, 1891.
Whittier, E. N., M.D., Boston, 1890, 1891.
Wilson, James C., A.M., M.D., Philadelphia, 1888, 1889, 1890, 1891.

Wirgman, Chas., M.D., Philadelphia, 1888.
Witherstine, C. Sumner, M.S., M.D., Philadelphia, 1888, 1889, 1890, 1891.
White, J. William, M.D., Philadelphia, 1889, 1890, 1891.
Young, Jas. K., M.D., Philadelphia, 1891.

JUNIOR ASSOCIATE EDITORS.

Baldy, J. M., M.D., Philadelphia, 1890.
Bliss, Arthur Ames, A.M., M.D., Philadelphia, 1889, 1891.
Cattell, H. W., M.D., Philadelphia, 1890, 1891.
Cerna, David, M.D., Ph.D., Philadelphia, 1891.
Clark, J. Payson, M.D., Boston, 1890, 1891.
Crandall, C. M., M.D., New York, 1891.
Cohen, Solomon Solis, A.M., M.D., Philadelphia, 1890, 1891.
Cryer, H. M., M.D., Philadelphia, 1889.
Deale, Henry B., M.D., Washington, 1891.
Dolley, C. S., M.D., Philadelphia, 1889, 1890, 1891.
Dollinger, Julius, M.D., Philadelphia, 1889.
Dorland, W. A., M.D., Philadelphia, 1891.
Freeman, Leonard, M.D., Cincinnati, 1891.
Goodell, W. Constantine, M.D., Philadelphia, 1888, 1889, 1890.
Gould, Geo. M., M.D., Philadelphia, 1889, 1890.
Greene, E. M., M.D., Boston, 1891.
Griffith, J. P. Crozer, M.D., Philadelphia, 1888.
Hoag, Junius, M.D., Chicago, 1888.
Howell, W. H., Ph.D., B.A., Baltimore, 1888, 1889.
Hunt, William, M.D., Philadelphia, 1888, 1889.
Jackson, Henry, M.D., Boston, 1891.
Kirk, Edward C., D.D.S., Philadelphia, 1888.
Lloyd, James Hendrie, M.D., Philadelphia, 1888.
McDonald, Willis G., M.D., Albany, 1889.
Penrose, Chas. B., M.D., Philadelphia, 1890.
Powell, W. M., M.D., Philadelphia, 1889, 1890, 1891.
Quinby, Chas. E., M.D., New York, 1889.
Sayre, Reginald H., M.D., New York, 1890, 1891.
Smith, Allen J., A.M., M.D., Philadelphia, 1889, 1890.
Vickery, H. F., M.D., Boston, 1891.
Warfield, Ridgely B., M.D., Baltimore, 1891.
Warner, Frederick M., M.D., New York, 1891.
Weed, Charles L., A.M., M.D., Philadelphia, 1888, 1889.
Wells, Brooks H., M.D., New York, 1888, 1889, 1890, 1891.
Wolff, Lawrence, M.D., Philadelphia, 1890.
Wyman, Walter, A.M., M.D., Washington, 1891.

ASSISTANTS TO ASSOCIATE EDITORS.

Baruch, S., M.D., New York, 1888.
Beatty, Franklin T., M.D., Philadelphia, 1888.
Brown, Dillon, M.D., New York, 1888.
Burcklater, A. F., M.D., New York, 1888.
Burr, Chas. W., M.D., Philadelphia, 1891.
Cohen, Solomon Solis, M.D., Philadelphia, 1889.
Cooke, R. G., M.D., New York, 1888.
Coolidge, Algernon, Jr., M.D., Boston, 1890.
Currier, A. F., M.D., New York, 1888.
Daniels, F. H., A.M., M.D., New York, 1888.
Deale, Henry B., M.D., Washington, 1890.
Eshner, A. A., M.D., Philadelphia, 1891.
Gould, George M., M.D., Philadelphia, 1888.
Granden, James H., M.D., New York, 1888, 1889.
Greene, E. M., M.D., Boston, 1890.
Guitéras, G. M., M.D., Washington, 1890.
Hance, I. H., A.M., M.D., New York, 1891.
Klingensmidt, C. H. A., M.D., Washington, 1890.
Martin, Edward, M.D., Philadelphia, 1891.
McKee, E. S., M.D., Cincinnati, 1889, 1890, 1891.
Myers, E. H., M.D., New York, 1888.
Packard, F. A., M.D., Philadelphia, 1890.
Pritchard, W. B., M.D., New York, 1891.
Sangree, E. B., A.M., M.D., Philadelphia, 1890.
Sears, G. G., M.D., Boston, 1890.
Shulz, R. C., M.D., New York, 1891.
Souwers, Geo. F., M.D., Philadelphia, 1888.
Taylor, H. L., M.D., Cincinnati, 1889, 1890.
Vansant, Eugene L., M.D., Philadelphia, 1888.

ASSISTANTS TO ASSOCIATE
EDITORS—(CONTINUED).

Vickery, H. F., M.D., Boston, 1880.
Warner, F. M., M.D., New York, 1888, 1889, 1890.
Wells, Brooks H., M.D., New York, 1888.
Wendt, E. C., M.D., New York, 1888.
Wilder, W. H., M.D., Cincinnati, 1889.
Wilson, C. Meigs, M.D., Philadelphia, 1889.
Wilson, W. R., M.D., Philadelphia, 1801.

CORRESPONDING STAFF.
EUROPE.

Antal, Dr. Gesa v., Buda-Pesth, Hungary.
Baginsky, Dr. A., Berlin, Germany.
Baratoux, Dr. J., Paris, France.
Barker, Mr. A. E. J., London, England.
Barnes, Dr. Fancourt, London, England.
Bayer, Dr. Carl, Prague, Austria.
Bouchut, Dr. E., Paris, France.
Bourneville, Dr. A., Paris, France.
Bramwell, Dr. Byrom, Edinburgh, Scotland.
Carter, Mr. William, Liverpool, England.
Caspari, Dr. G. A., Moscow, Russia.
Chirali y Selma, Dr. V., Seville, Spain.
Cordes, Dr. A., Geneva, Switzerland.
D'Estrees, Dr. Debout, Contrexéville, France.
Diakonoff, Dr. P. J., Moscow, Russia.
Dobrashian, Dr. G. S., Constantinople, Turkey.
Doléris, Dr. L., Paris, France.
Doutrelepont, Prof., Bonn, Germany.
Doyon, Dr. H., Lyons, France.
Drzewiecki, Dr. Jos., Warsaw, Poland.
Dubois-Reymond, Prof., Berlin, Germany.
Ducrey, Dr. A., Naples, Italy.
Dujardin-Beaumetz Dr., Paris, France.
Duke, Dr. Alexander, Dublin, Ireland.
Eklund, Dr. F., Stockholm, Sweden.
Fokker, Dr. A. P., Groningen, Holland.
Fort, Dr. J. A., Paris, France.
Fournier, Dr. Henri, Paris, France.
Franks, Dr. Kendal, Dublin, Ireland.
Frémy, Dr. H., Nice, France.
Fry, Dr. George, Dublin, Ireland.
Golowina, Dr. A., Varna, Bulgaria.
Gouguenheim, Dr. A., Paris, France.
Haig, Dr. A., London, England.
Hamon, Mr. A., Paris, France.
Harley, Mr. V., London, England.
Harley, Mr. H. R., Nottingham, England.
Harley, Prof. Geo., London, England.
Harpe, Dr. de la, Lausanne, Switzerland.
Hartmann, Prof. Arthur, Berlin, Germany.
Heitzmann, Dr. J., Vienna, Austria.
Helferich, Prof., Greifswald, Germany.
Hewetson, Dr. Bendelack, Leeds, England.
Hoff, Dr. E. M., Copenhagen, Denmark.
Humphreys, Dr. F. Rowland, London, England.
Illingworth, Dr. C. K., Accrington, England.
Jones, Dr. D. M. de Silva, Lisbon, Portugal.
Knott, Dr. J. F., Dublin, Ireland.
Krause, Dr. H., Berlin, Germany.
Landolt, Dr. E., Paris, France.
Levison, Dr. J., Copenhagen, Denmark.
Lutaud, Dr. A., Paris, France.
Mackay, Dr. W. A., Huelva, Spain.
Mackowen, Dr. T. C., Capri, Italy.
Manché, Dr. L., Valetta, Malta.
Massei, Prof. F., Naples, Italy.
Mendez, Prof. R., Barcelona, Spain.
Meyer, Dr. E., Naples, Italy.
Meyer, Prof. W., Copenhagen, Denmark.
Monod, Dr. Charles, Paris, France.
Montefusco, Prof. A., Naples, Italy.
More-Madden, Prof. Thomas, Dublin, Ireland.
Morel, Dr. J., Ghent, Belgium.
Mygind, Dr. Holger, Copenhagen, Denmark.
Mynlieff, Dr. A., Breukelen, Holland.
Napier, Dr. A. D. Leith, London, England.
Nicolich, Dr., Trieste, Austria.
Oberländer, Dr., Dresden, Germany.
Obersteiner, Prof., Vienna, Austria.
Pampoukis, Dr., Athens, Greece.
Pansoni, Dr., Naples, Italy.
Parker, Mr. Rushton, Liverpool, England.
Pel, Prof. P. K., Amsterdam, Holland.
Pippinskjold, Dr., Helsingfors, Finland.
Pulido, Prof. Angel, Madrid, Spain.

Róna, Dr. S., Buda-Pesth, Hungary.
Rosenbusch, Dr. L., Lvov, Galicia.
Rossbach, Prof. M. F., Jena, Germany.
St. Germain, Dr. de, Paris, France.
Sänger, Prof. M., Leipzig, Germany.
Santa, Dr. P. de Pietra, Paris, France.
Schiffers, Prof., Liége, Belgium.
Schmiegelow, Prof. E., Copenhagen, Denmark.
Scott, Dr. G. M., Moscow, Russia.
Simon, Dr. Jules, Paris, France.
Sollier, Dr. P., Paris, France.
Solowieff, Dr. A. N., Lipetz, Russia.
Sota, Prof. R. de la, Seville, Spain.
Sprimont, Dr., Moscow, Russia.
Stockvis, Prof. B. J., Amsterdam, Holland.
Szadek, Dr. Carl, Kiew, Russia.
Tait, Mr. Lawson, Birmingham, England.
Thiriar, Dr., Brussels, Belgium.
Trifiletti, Dr., Naples, Italy.
Tuke, Dr. D. Hack, London, England.
Ulrik, Dr. Axel, Copenhagen, Denmark.
Unverricht, Prof., Jena, Germany.
Van der Mey, Prof. G. H., Amsterdam, Holland.
Van Leent, Dr. F., Amsterdam, Holland.
Van Millingen, Prof. E., Constantinople, Turkey.
Van Rijnberk, Dr., Amsterdam, Holland.
Wilson, Dr. George, Leamington, England.
Wolfenden, Dr. Norris, London, England.
Zweifel, Prof., Leipzig, Germany.

AMERICA AND WEST INDIES.

Bittencourt, Dr. J. C., Rio Janeiro, Brazil.
Cooper, Dr. Austin S., Buenos Ayres, Argentine Republic.
Daguine, Prof. Manuel, Caracas, Venezuela.
Desvernine, Dr. C. M., Havana, Cuba.
Fernandez, Dr. J. L., Havana, Cuba.
Finlay, Dr. Charles, Havana, Cuba.
Fontecha, Prof. R., Tegucigalpa, Honduras.
Harvey, Dr. Eldon, Hamilton, Bermuda.
Herdocia, Dr. E. Leon, Nicaragua.
Levi, Dr. Joseph, Colon, U. S. Columbia.
Mello, Dr. Vierra de, Rio Janeiro, Brazil.
Moir, Dr. J. W., Belize, British Honduras.
Moncorvo, Prof., Rio Janeiro, Brazil.
Plá, Dr. E. F., Havana, Cuba.
Rake, Dr. Beaven, Trinidad.
Rincon, Dr. F., Maracaibo, Venezuela.
Semeleder, Dr. F., Mexico, Mexico.
Soriano, Dr. M. S., Mexico, Mexico.
Strachan, Dr. Henry, Kingston, Jamaica.

OCEANICA, AFRICA, AND ASIA.

Baelz, Prof. B., Tokyo, Japan.
Barrett, Dr. Jas. W., Melbourne, Australia.
Branfoot, Dr. A. M., Madras, India.
Carageorgiades, Dr. J. G., Limassol, Cyprus.
Cochran, Dr. Joseph P., Oroomiah, Persia.
Coltman, Dr. Robert, Jr., Che-foo, China.
Condict, Dr. Alice W., Bombay, India.
Creece, Dr. John M., Sydney, Australia.
Dalzell, Dr. J., Umsiga, Natal.
Diamantopulos, Dr. Geo., Smyrna, Turkey.
Drake-Brockman, Dr. Madras, India.
Fitzgerald, Mr. T. N., Melbourne, Australia.
Foreman, Dr. L., Sydney, Australia.
Gabizaghian, Dr. Ohan, Adana, Asia Minor.
Grant, Dr. David, Melbourne, Australia.
Johnson, Dr. R., Dera Ishmail Khan, Beloochistan.
Kimura, Prof. J. K., Tokyo, Japan.
Knaggs, Dr. S., Sydney, Australia.
Manasseh, Dr. Beshara I., Brummana, Turkey in Asia.
McCandless, Dr. H. H. Hainan, China.
Moloney, Dr. J., Melbourne, Australia.
Neve, Dr. Arthur, Bombay, India.
Perez, Dr. George V., Puerto Orotava, Teneriffe.
Reid, Dr. John, Melbourne, Australia.
Robertson, Dr. W. S., Port Said, Egypt.
Rouvier, Prof. Jules, Beyrouth, Syria.
Scranton, Dr. William B., Seoul, Corea.
Sinclair, Dr. H., Sydney, Australia.
Thompson, Dr. James B., Petchaburee, Siam.
Wheeler, Dr. P. d'E., Jerusalem, Palestine.
Whitney, Dr. H. T., Foochow, China.
Whitney, Dr. W. Norton, Tokyo, Japan.

RANNEY

Lectures on Nervous Diseases.

FROM THE STAND-POINT OF CEREBRAL AND SPINAL LOCALIZATION, AND
THE LATER METHODS EMPLOYED IN THE DIAGNOSIS AND
TREATMENT OF THESE AFFECTIONS.

By AMBROSE L. RANNEY, A.M., M.D., Professor of the Anatomy and Physiology of the Nervous System in the New York Post-Graduate Medical School and Hospital; Professor of Nervous and Mental Diseases in the Medical Department of the University of Vermont, etc.; Author of "The Applied Anatomy of the Nervous System," "Practical Medical Anatomy," etc., etc.

It is now generally conceded that the nervous system controls all of the physical functions to a greater or less extent, and also that most of the symptoms encountered at the bedside can be explained and interpreted from the stand-point of nervous physiology.

Profusely illustrated with original diagrams and sketches in color by the author, carefully selected wood-engravings, and reproduced photographs of typical cases. One handsome royal octavo volume of 780 pages.

SOLD ONLY BY SUBSCRIPTION, OR SENT DIRECT ON RECEIPT OF PRICE.
SHIPPING EXPENSES PREPAID.

Price, in United States, Cloth, $5.50; Sheep, $6.50; Half-Russia, $7.00.
Canada (duty paid), Cloth, $6.05; Sheep, $7.15; Half-Russia, $7.70.
Great Britain, Cloth, 32s.; Sheep, 37s. 6d.; Half-Russia, 40s. France,
Cloth, 34 fr. 70; Sheep, 40 fr. 45; Half-Russia, 43 fr. 30.

We are glad to note that Dr. Ranney has published in book form his admirable lectures on nervous diseases. His book contains over seven hundred large pages, and is profusely illustrated with original diagrams and sketches in colors, and with many carefully selected woodcuts and reproduced photographs of typical cases. A large amount of valuable information, not a little of which has but recently appeared in medical literature, is presented in compact form, and thus made easily accessible. In our opinion, Dr. Ranney's book ought to meet with a cordial reception at the hands of the medical profession, for, even though the author's views may be sometimes open to question, it cannot be disputed that his work bears evidence of scientific method and honest opinion.—*American Journal of Insanity.*

STANTON'S

Practical and Scientific Physiognomy;

OR

How to Read Faces.

By MARY OLMSTED STANTON. Copiously illustrated. Two large Octavo volumes.

The author, MRS. MARY O. STANTON, has given over twenty years to the preparation of this work. Her style is easy, and, by her happy method of illustration of every point, the book reads like a novel and memorizes itself. To physicians the diagnostic information conveyed is invaluable. To the general reader each page opens a new train of ideas. (This book has no reference whatever to phrenology.)

SOLD ONLY BY SUBSCRIPTION, OR SENT DIRECT ON RECEIPT OF PRICE.
SHIPPING EXPENSES PREPAID.

Price, in United States, Cloth, $9.00; Sheep, $11.00; Half-Russia, $13.00.
Canada (duty paid), Cloth, $10.00; Sheep, $12.10; Half-Russia,
$14.30. Great Britain, Cloth, 56s.; Sheep, 68s.; Half-Russia, 80s.
France, Cloth, 30 fr. 30; Sheep, 36 fr. 40; Half-Russia, 43 fr. 30.

SAJOUS

Lectures on the Diseases of the Nose and Throat.

DELIVERED AT THE JEFFERSON MEDICAL COLLEGE, PHILADELPHIA.

By CHARLES E. SAJOUS, M.D. Formerly Lecturer on Rhinology and Laryngology in Jefferson Medical College; Vice-President of the American Laryngological Association; Officer of the Academy of France and of Public Instruction of Venezuela; Corresponding Member of the Royal Society of Belgium, of the Medical Society of Warsaw (Poland), and of the Society of Hygiene of France; Member of the American Philosophical Society, etc., etc.

☞ *Since the publisher brought this valuable work before the profession, it has become: 1st, the text-book of a large number of colleges; 2d, the reference-book of the U. S. Army, Navy, and the Marine Service; and, 3d, an important and valued addition to the libraries of over 10,000 physicians.*

This book has not only the inherent merit of presenting a clear *exposé* of the subject, but it is written with a view to enable the general practitioner to treat his cases himself. To facilitate diagnosis, *colored plates* are introduced, showing the appearance of the different parts in the diseased state as they appear in nature by artificial light. No error can thus be made, as each affection of the nose and throat has its representative in the *100 chromo-lithographs* presented. In the matter of treatment, the indications are so complete that even the slightest procedures, folding of cotton for the forceps, the use of the probe, etc., are clearly explained.

Illustrated with 100 chromo-lithographs, from oil paintings by the author, and 93 engravings on wood. One handsome royal octavo volume.

SOLD ONLY BY SUBSCRIPTION, OR SENT DIRECT ON RECEIPT OF PRICE, SHIPPING EXPENSES PREPAID.

Price, in United States, Cloth, Royal Octavo, $4.00; Half-Russia, Royal Octavo, $5.00. Canada (duty paid), Cloth, $4.40; Half-Russia, $5.50. Great Britain, Cloth, 22s. 6d.; Sheep or Half-Russia, 28s. France, Cloth, 24 fr. 60; Half-Russia, 30 fr. 30.

It is intended to furnish the general practitioner not only with a guide for the treatment of diseases of the nose and throat, but also to place before him a representation of the normal and diseased parts as they would appear to him were they seen in the living subject. As a guide to the treatment of the nose and throat, we can cordially recommend this work. —*Boston Medical and Surgical Journal.*

IN PRESS.

THE CHINESE: Their Present and Future; Medical, Political, and Social.

By ROBERT COLTMAN, Jr., M.D., Surgeon in Charge of the Presbyterian Hospital and Dispensary at Teng Chow Fu; Consulting Physician of the American Southern Baptist Mission Society; Examiner in Surgery and Diseases of the Eye for the Shantung Medical Class; Consulting Physician to the English Baptist Missions, etc. Illustrated with about Sixteen Fine Engravings from photographs of persons, places, and objects characteristic of China. In one Octavo volume of about 250 pages. READY ABOUT DECEMBER 1, 1891.

The author has spent many years among the Chinese; LIVED with them in their dwellings; THOROUGHLY learned the language; has become conversant with all their strange and odd characteristics to a greater extent than almost any other American. He has been a physician to all classes of this wonderful people, and the opportunities thus afforded for a clear insight into the inner life of the Chinese, their virtues and vices, political, social, and sanitary condition, probable destiny, and their present important position in the world to-day, have been ably and wisely used by Dr. Coltman.

Nearly Ready. Will be Issued about October 1, 1891.

Age of the Domestic Animals.

BEING A COMPLETE TREATISE ON THE DENTITION OF THE HORSE, OX,
SHEEP, HOG, AND DOG, AND ON THE VARIOUS OTHER MEANS
OF TELLING THE AGE OF THESE ANIMALS.

By RUSH SHIPPEN HUIDEKOPER, M.D., Veterinarian, Alfort, France.
Professor of Sanitary Medicine and Veterinary Jurisprudence in the
American Veterinary College, New York; Lieutenant-Colonel and Sur-
geon-in-Chief National Guard of Pennsylvania; Fellow of the College
of Physicians, Philadelphia; Honorary Fellow of the Royal College
Veterinary Surgeons, London; Late Dean of the Veterinary Department
University of Pennsylvania, etc., etc.

Complete in one handsome Royal Octavo volume, with about 160
Illustrations. This is one of the most important works on the domestic
animals published in recent years.

Ready Very Shortly.

A, B, C of the Swedish System of Educational Gymnastics.

A PRACTICAL HAND-BOOK FOR SCHOOL-TEACHERS AND THE HOME.

By HARTVIG NISSEN, Instructor of Physical Training in the Public
Schools of Boston, Mass.; Instructor of Swedish and German Gymnas-
tics at Harvard University's Summer School, 1891; Author of "A
Manual on Swedish Movement and Massage Treatment," etc., etc.

Complete in one neat 12mo volume, bound in extra flexible cloth
and appropriately illustrated with 77 excellent engravings aptly eluci-
dating the text.

Lectures on Auto-Intoxication.

By PROF. BOUCHARD, Paris. Translated from the French, with an
Original Appendix by the author. By THOMAS OLIVER, M.D., Professor
of Physiology in University of Durham, England. IN PRESS.

www.ingramcontent.com/pod-product-compliance
Lightning Source LLC
Chambersburg PA
CBHW021521210326
41599CB00012B/1339